W9-AXH-615

Air Vagabonds

Air Vagabonds

Oceans, Airmen, and a Quest for Adventure

Anthony J. Vallone

Smithsonian Books
Washington and London

Editor: Robert A. Poarch
Designer: Brian Barth

Library of Congress Cataloging-in-Publication Data
Vallone, Anthony J.
Air vagabonds : oceans, airmen, and a quest for adventure / Anthony J. Vallone.
p. cm.
ISBN 1-58834-137-2 (alk. paper)
1. Vallone, Anthony J. 2. Air pilots—United States—Biography.
3. Airplanes—Ferrying. I. Title.
TL540.V28 V35 2003
629.13′092—dc21
[B] 2002042614

British Library Cataloguing-in-Publication Data is available

Manufactured in the United States of America
10 09 08 07 06 05 04 03 5 4 3 2 1

∞The paper used in this publication meets the minimum requirements of the American National Standard for Information Sciences—Permanence of Paper for Printed Library Materials ANSI Z39.48-1984.

To those ocean ferry pilots who are with the Trim God

There's a race of men that don't fit in,
A race that can't stay still;
So they break the hearts of kith and kin,
And they roam the world at will.
They range the field and they rove the flood,
and they climb the mountain's crest;
Theirs is the curse of the gypsy blood,
And they don't know how to rest.
—from *The Men That Don't Fit In* by Robert Service

"You can't send a kid up in a crate like this."
—from the screenplay *East of Samoa*

Contents

Acknowledgments ix
List of Acronyms and Abbreviations xi
Introduction xiii

1 Encounter with Fate 1
2 The Call of Adventure 10
3 Earning My Wings . . . across the Pacific 19
4 The Route to Singapore 35
5 Teaching the Ropes 53
6 The North Atlantic 65
7 Land of the Midnight Sun 79
8 A Ship without a Rudder 97
9 Into the Revolution 113
10 A Strange Coincidence 132
11 The Making of a Guru 145

12 An Albatross to the Far East 160

13 Escape 184

14 The Dark Continent 199

15 Adrift in the Kingdom 220

16 The Boom Years 238

17 Farewell from the Hereafter 260

18 Merchants of Counterinsurgency 272

19 Return to the Fold 286

Acknowledgments

The writing of this book could not have been possible without the help of friends, family, and fellow pilots too numerous to include here. For that glaring omission, I ask not for their forgiveness, but at least their understanding. I would be remiss, however, if I neglected to mention those individuals without whose help this book would be nothing more than a project unrealized or a paperweight sitting on my bookshelf in manuscript form.

To start at the beginning, I wish to thank my good friend Gert Norton, who listened so often to my ideas about someday writing a book about ferrying airplanes that she finally told me to quit talking about it and just write it. For this, as well as her tireless critiquing of draft after endless draft, I am eternally grateful.

My gratitude also goes to Sandy Moyle, who led me to discover the dying art of writing letters and encouraged me during the early stages of the manuscript's development.

Thank you to my agent Kristen Wainwright and her staff, whose commitment to seeing the book in print went beyond any expectations of monetary reward. And also my editor Mark Gatlin, who recognized the importance that the little-known adventure of ferrying small airplanes over the ocean played in the history of aviation.

For help with research and in locating many of the pilots I had flown with, I'm indebted to two people: Donna Waldman, who kept in touch with pilots whose later careers had scattered them to the four winds, and Donn Kerby, for so often digging through the archival rubble of his apartment to find an old flight chart or any notes he may have written about something that happened a quarter century ago. Of the sixty-three people mentioned in this book, it is inevitable there would be some whose present whereabouts are shrouded behind the veils of geography and time. Wherever fortune has landed them, my thanks go to George Rice, Bob Campbell, Lev Flournoy, Paul Crandall, Peter Goldstern, Andy Mattenheimer, Pat White, Ernst Krupp, Molly Willett, and Gus Grillo. I've attempted to portray them and the events in which they participated accurately, and I hope, for the most part, that they will be pleased.

Lastly, my deepest thanks go to all the ferry pilots, the mechanics, the airport workers, and their wives who were part of this forgotten era and who's adventures make up the brunt of this book. Though decades may have past since we last spoke, they were always ready to swap stories—and undoubtedly a few lies—in fond remembrance of a time unlikely to ever return.

Acronyms and Abbreviations

ADF	automatic direction finder (navigation radio)
ANC	African National Congress (South African political party)
ATC	air traffic control
BBC	British Broadcasting Corporation
BOAC	British Overseas Airways Corporation (airline)
BOQ	bachelor officer's quarters
CAB	Civil Aeronautics Board (U.S. air carrier regulating agency)
CIA	Central Intelligence Agency
FAA	Federal Aviation Administration
FBI	Federal Bureau of Investigation
FBO	fixed base operation (aircraft service facility)
FNLA	Frente Nacional para a Libertaçäo de Angola (National Front for the Liberation of Angola—Rebel group based in Zaire, backed by Zaire, China, and briefly by the United States)
GPS	global positioning system (satellite navigation system)
HF	high frequency (long-range radio communication system)
ICAO	International Civil Aviation Organization

ILS instrument landing system (instrument approach navigation system)
IRA Irish Republican Army
JFK John Fitzgerald Kennedy International Airport in New York City
KGB Komitet Gosudarstyennoy Bezopasnosti (Committee for State Security—Soviet Union intelligence directorate)
KLM Royal Dutch Airlines
LORAN long range navigation (navigation system)
MOT Ministry of Transportation (Canadian air transportation regulating department)
MPLA Movimento Popular de Libertaçäo de Angola (Popular Movement for the Liberation of Angola—communist-backed rebel group)
NDB non-directional beacon (low- and medium-frequency navigation radio)
OPEC Organization of Petroleum Exporting Countries
PSA Pacific Southwest Airlines
RCAF Royal Canadian Air Force
RCMP Royal Canadian Mounted Police
SAS Scandinavian Airlines System
SEC Securities and Exchange Commission
SWAPO South West African People's Organization (Namibian political party)
TAP Transportadora Aerea Portuguesa (Portuguese airline)
UNITA Uniäo Nacional para a Independênciá Total de Angola (National Union for the Total Independence of Angola—U.S. and South African–backed rebel group)
VHF very high frequency (radio frequency band)
VOR variable omni range (VHF navigation radio)

Introduction

Far out over the North Atlantic, closer to the realm of mariners and sea creatures than to jet aircraft and the airmen who fly them, a small, piston-engine airplane plods along a mile or so above the waves. A hundred thoughts race through the pilot's mind as he whiles away the hours until once again he is safely over solid earth. *Will the airplane pick up ice in the approaching cold front? Will the temporary long-range fuel tanks start leaking? Will the engine keep running? Were the forecast winds accurate, or will the airplane be blown hundreds of miles off course?* Any of these problems, if they were to develop, could force the pilot to ditch into an unforgiving ocean.

The demons that could abruptly end a ferry pilot's life are infinite. Yet, during the halcyon days of the 1970s and 1980s, when production of small aircraft was at levels never before attained in the history of general aviation, at any given time and over any of the Earth's oceans a small airplane was being ferried to its new owners somewhere beyond the horizon.

The events chronicled in this book tell the story of a small band of intrepid aviators who were in the cockpits of those aircraft, battling the cold, the sweltering heat, and the trigger-happy third-world armies that shot first and asked questions later. As one of those pilots, I was privileged to know many of the men and women whose exploits fill these pages. I am lucky to have survived the same adventures that took the lives of my less fortunate comrades.

We all took the same chances and flew the same type of airplanes over the same routes and in the same weather. Yet, each year, on average, three out of less than a hundred professional ferry pilots throughout the world took off and were never seen again. In an attempt to find logic in the arbitrary selection of who lived and who died, and why at times our flights went so smoothly it seemed we were surely blessed, while at other times we suffered all the misfortunes of having been born under a malignant star, we imagined such destiny to be the workings of a deity.

Being neither inordinately religious nor of sufficient faith to stand passively by while our fate was determined by something so nebulous as luck or random chance, we ascribed to a belief that our fortunes were dependent solely on the disposition of the Trim God.

Trim, as in the trim tabs on an aircraft's flight controls, refers to the aerodynamic harmony between airplane and slipstream. If the trim tabs are not adjusted properly—"out of trim" in pilot jargon—drag is induced, and the airplane will fly more slowly and mush through the air in defiance of the laws of streamlined flight. As the unseen force affecting the speed at which an airplane flies, trim is considered a cause for both good luck and bad.

In pagan cultures gods demanded some form of tribute, whether simple observance or fealty, gold or silver, or the ritual sacrifice of newborn infants or vestal virgins. Our creation of the Trim God was no different. Although we had no formal doctrine decreeing what those observances should be, we believed this deity to be the god of preparedness and caution. If a pilot heeded these tenets, he would be granted tailwinds and his trip would be speedy and problem-free. But if he was careless or took unnecessary risks, he would be cursed with headwinds, lousy weather, broken airplanes, and one disastrous problem after another that, depending on the severity of the offense, would be dealt the ultimate penance.

The origin of the Trim God is forgotten, but is most likely attributable to no one individual or event—it just sort of evolved into more frequent usage in our conversation. "He must have angered the Trim God," we might have said when a pilot suffered a number of problems; or someone would say, "He's gone to the Trim God," when one of us did not return. Finding some measure of solace in the explanation of mysteries for which there were no answers, we went about our work, flying one trip after another, never knowing if our next flight would be the one where we would be summoned by the Trim God.

The Trim God is no longer as active as it once was. Neither is general aviation. With the demise of small-aircraft manufacturing, ferrying has been reduced to a mere footnote in the annals of international commerce and aviation legend. As for the pilots, both living and dead, their stories can be heard in operations offices and crew lounges from Gander to Singapore. Or one can meander through the following pages for a firsthand chronicle of this bygone era—a time before global positioning systems when small airplanes were navigated across the oceans by wind and compass, before deregulation of the airlines when the Clipper Ships of Pan American World Airways ruled the skies and the old world order of two superpowers dominated the globe, when Europe still maintained influence over its African colonies and flying across borders was often a harrowing adventure.

Though written from my perspective, this book is not about me. It is a tale of a group of airmen who knowingly accepted the dangers and loneliness of their profession so that they might find solace in an unfettered life of wandering around the world.

This is their story.

1

Encounter with Fate

Parked under the wing of the German Air Force C-130 Hercules, the Piper Cadet looked like a scaled-down mock-up of a real airplane. It was almost small enough to fit inside one of the cargo plane's four engines. Short and stubby, the little airplane gave no impression of speed, let alone comfort. It was simply a trainer: one engine, two seats, and a minimum of avionics. This was Piper's answer to the skyrocketing inflation of the 1980s, a basic bare-bones airplane that sold for under sixty thousand dollars. The Cadet was one of the smallest airplanes I had ever ferried across the ocean. Even the name suggested it wasn't full grown.

I checked off the coordinates as the control tower operator read back my clearance for the route from St. Johns, Newfoundland, to Santa Maria in the Azores. Tomorrow night I would make the run over to Nantes, on the French coast, before the Atlantic Ocean and all of its dangers would be behind me. I had flown the route so often, it was like a milk run—yet, one that still demanded respect and caused many an hour of anxiety.

Between Newfoundland and the Azores, a small group of islands in the southern part of the North Atlantic, lie fourteen hundred miles of ocean that over the years has claimed the lives of scores of pilots. In the few dozen times I had flown the route, I had a few close calls myself, but always managed to arrive at Santa Maria safely. It was almost becoming routine. This time, however, I couldn't shake this nagging feeling that it would not be routine. I had ferried half a dozen Cadets to Europe over the past few years, but never across the ocean in winter.

It was a humbling experience. The overcast sky and banks of drifting snow added to that a feeling of danger I could just as well have done without. I'm certain the controller felt sorry for me as I taxied to the runway, alone, in the dead of night.

"November-Nine-Zero-Foxtrot, from St. Johns Tower. You're cleared for takeoff," said the controller, adding a somber, "And good luck."

A light snow had begun to fall. At full power, the Cadet accelerated slowly at first, as if in protest to the ridiculous weight it was being asked to carry aloft—one hundred fifteen gallons of extra fuel stored in three internal tanks strapped to the floor beside and behind me. The airplane used more than half of the seven-thousand-foot runway before gaining enough speed to support itself in flight. I eased back on the control wheel, and the airplane lifted into the air, at first seeming to hover, its speed over the ground impeded by the strong headwind. Runway lights crept beneath me in the darkness and then disappeared as airspeed increased and the airplane began a slow climb.

Like the other times I had flown a Cadet, I spent the first five minutes of the climb wondering what I had forgotten. While most airplanes have an after-takeoff checklist with a dozen or more items the pilot has to comply with—adjust power, propellers, mixture, fuel pumps, cowl flaps, etc.—the Cadet had only one: turn off the boost pump. Although I complied with this simple instruction, I still couldn't shake the feeling that I had forgotten something that would come back to haunt me.

After two hundred and sixty ocean crossings, I still felt uneasy every time I headed out over open water. It was worse on nights like tonight, when low clouds and snow blocked any glimmer of light from moon or stars. All was dark, like flying into a black bag that becomes darker and lonelier the farther I got from St. Johns. It was just one of the dangers of ferrying small airplanes over the ocean, where vast distances and slow speeds—especially during the dwindling daylight hours of winter—meant at least part of the leg had to be

Sondrestrom
Akureyri
Umea
Reykjavik
Narsarsuaq
Shannon
Goose Bay
Gander
Kassel
St Johns
Tashkent
San Francisco
Lock Haven
Lisbon
Santa Maria
Seoul
Tokyo
Lakeland
Tenerife
Tripoli
Cairo
Karachi
Honolulu
El Aiun
Luxor
Bahrain
Jeddah
Bombay
Hong Kong
Taiwan
Wake Is.
Kano
Dakar
Manila
Khartoum
Bangkok
Guam
Abidjan
Caracas
Christmas Is.
Nairobi
Majuro
Sao Tome
Libreville
Singapore
Truk
Tarawa
Dar es Salaam
Seychelles
Pago Pago
Luanda
Jakarta
Malang
Solomon Is.
Tahiti
Windhoek
Darwin
Fiji Is.
Mount Isa
New Caledonia
Johannesburg
Brisbane
Norfolk Is.
Perth
Sydney
Auckland

The world

flown at night. Although I didn't like taking off at night in bad weather, it was either that or risk getting stuck for several days. The weather system approaching Newfoundland was forecast to get worse, and none of the briefers would even guess as to when it might move out to sea. If there was one thing I had learned from ferrying over the years, it was that you had to take chances.

I was climbing through nine hundred feet when St. Johns Tower directed me to turn onto a southeasterly course toward Santa Maria and then handed me off to Gander Approach Control just as I entered the overcast. The clouds were much lower than the weather briefer had forecast, which caused me to wonder if he might also have been wrong about the altitude of the tops of the clouds.

Five thousand feet, he had said. I hoped he was right, because as slow as the airplane was climbing, I figured it would take fifteen minutes to reach that altitude—if I didn't pick up any ice in the clouds. With the outside temperature below zero, any moisture I fly into would freeze as soon as it hit the airplane. I hadn't picked up any yet, but with snow you could never be certain. If there was one thing I was certain about, it was that the Cadet wouldn't carry much ice. The airplane just wasn't designed for long ocean flights, not in the summer and especially not in the winter. Slow and underpowered under the best of conditions, with the weight of the extra fuel it was even more sluggish. It crept through the dense air inch by inch, each mile flown a painstakingly small investment toward my final destination—Kassel, Germany—still a hemisphere away.

The mental ritual of preparing for the crossing and worrying about all the things that could go wrong was the same every time, at least during the climbout. Once I reach cruising altitude and trim the airplane for the long ride, then I can relax, for I'll have reached the first milestone of the flight. I knew from the other times I had flown a Cadet that it takes thirty minutes or so to reach that milestone, which for tonight's crossing was an altitude of nine thousand feet. During that time, there wasn't much to do but wait—and hope the Trim God was in a good mood.

However favorable my standing with this alleged deity may have been on previous crossings, it looked as if it would not be enough tonight. When I looked out the side window I saw a light trace of ice forming on the leading edge of the wing. It wasn't much, but on a Cadet, even a little ice was too much. The airplane began losing speed and climbing slower. As I penetrated deeper into the clouds, the ice thickened and speed started falling off rapidly.

After what seemed like hours, the airplane finally reached five thousand feet. But I was still in clouds, and ice was still building. The rough, clear ice that formed in jagged chunks disrupted the airflow over the wings so much that the airplane struggled to climb higher. But I had to keep climbing. I had to get above the clouds.

How much higher would I have to go? A thousand feet? Two thousand? Already the tops were higher than forecast, and it was so dense at times that I could barely see the wingtips, let alone an occasional star peaking through the haze that might indicate I was close to the top of the overcast. Even if the tops were a hundred feet above me, it would still be too high, because at fifty-seven hundred feet the Cadet had given all it could and refused to climb another twenty feet. That wasn't the only thing I had to worry about. I had been so preoccupied with the ice that I didn't even think about the winds. I realize now that they, too, were worse than forecast. I had been in the air fifteen minutes, yet was only twenty-one miles out of St. Johns. That worked out to less than seventy knots over the ground. At that speed, I would run out of fuel long before I reached the Azores.

Time was rapidly approaching when I would have to decide whether or not to return to St. Johns. Maybe it was admitting defeat, or maybe in some twisted logic I just didn't want to lose what little progress I had made so far, but I delayed making that decision. Any minute now I expected to break out into a clear sky and would then feel foolish for having entertained thoughts of aborting the flight when salvation was so near at hand. After all, I had been in situations like this before and always managed to get out of them. Why should tonight be any different?

Still, I knew I had to set a limit on how long I would continue. I promised myself that if I didn't break out of the clouds in the next few minutes I would return to St. Johns. But I neglected to set a specific time, because I was certain I would be in clear air before it got too serious. So I kept going farther from the airport, watching the ice build and my airspeed deteriorate, yet fully prepared to turn back if the weather didn't improve in the next few minutes.

How many next few minutes passed, I've no idea. The only thing I did know was that I should have turned around sooner, because the airplane now had so much ice on its wings that it was starting to lose altitude. The descent was only about a hundred feet per minute, nowhere near enough to cause panic, but with no break in the clouds, I knew it would get worse. Reluctantly, I banked the airplane into a one-hundred-eighty-degree turn and called Gander Approach to let them know I was returning to St. Johns.

"Will you be able to maintain five thousand feet?" the controller asked.

"Negative, Gander. I'm picking up too much ice. I need you to clear all altitudes below me." Ice was still building, and I was now losing altitude at the rate of three hundred feet per minute. Three hundred into forty-five hundred feet—my altitude at this exact second—came out to fifteen minutes before I would run out of altitude. That wouldn't be too bad, except that I was eighteen minutes from the airport.

I pondered the arithmetic in disbelief. It couldn't be right. I must have made a mistake. I recalculated the figures, and, when still they showed no change, I tried again. But no matter how many times I worked the numbers, I kept coming up three minutes short.

"November-Nine-Zero-Foxtrot, I show you twenty-two miles from St. Johns Airport. Confirm you'll be able to land there." The controller, one hundred miles away in Gander and clear of the storm, probably wasn't prepared for my answer.

"It doesn't look that way, Gander. I'm losing altitude at three hundred feet per minute, and the arithmetic isn't coming out in my favor," I said.

The radio was silent for several seconds before the controller's steady voice came back over the speaker. "Okay, November-Nine-Zero-Foxtrot, you're cleared direct to St. Johns Airport to maintain altitude at your discretion. Runway three-four is active, and you're cleared for a straight-in approach. We've notified search and rescue, and they're standing by."

I couldn't get over how calm and reassuring the faceless voice of the controller was. I tried to picture the darkened control room as this stranger watched a tiny electronic blip of light move slowly across the radar screen. His voice never wavered or cracked, and for that I was thankful, for although I appeared calm, it was mostly because I was still just reacting to the situation and hadn't had time to think about it. The slightest tremor in the controller's voice, however, would have been enough to cause me to panic.

It was now that the realization hit me, not like a bolt of lightning, but just kind of sneaking up on me. There would be no escape tonight. I had been careless, and now I was going to pay for it. There was no other way this flight could possibly end except in a flaming pile of rubble, the airplane twisted beyond recognition after plowing into trees or telephone poles or buildings I wouldn't see until it was too late. I'm certain the shameful tone of fear had already crept into my voice and raised it an octave or two no matter how steady I tried to sound. But I had to keep it under control. Even if one were rushing headlong toward imminent death, aviation tradition dictated a calm, almost detached manner. I knew that if I panicked even a little, I might as well abandon any hope of sound judgment. And the one thing I needed now was to be able to make logical decisions.

If the clouds weren't any lower than when I took off, I should break out in time to see the lights of the city and avoid crashing into a populated area. A road might be a good place to set down, but in the darkness I risked hitting a high-tension line that would either electrocute me on the spot or send the airplane tumbling end over end. My best chance would be to aim for a dark area and hope that if there were trees, I would land on top of them and not fly into them.

The ice on the wings looked about three-quarters of an inch thick, and, although I was still in clouds, it didn't appear to be getting thicker. As fuel burned off and the airplane became lighter, I was no longer losing altitude as fast as I had earlier. Even with that little help I figured I would still land about a mile short of the airport.

"November-Nine-Zero-Foxtrot, I show you at twenty-eight hundred feet and sixteen miles. Be advised of the eleven-hundred-foot hills just southeast of the airport," the controller warned.

With everything else going on, I had completely forgotten about the hills between my present position and the airport. I would be a few hundred feet below their peaks when I approached them. The bases of the clouds were at nine hundred feet when I took off, and if they hadn't lifted, they would obscure the tops of the hills. This gave me something else to worry about, for while I might be willing to take my chances with a crash landing on level ground, trying to avoid a hillside while flying in clouds would be suicide. There was only one thing to do—turn toward open water and ditch in the ocean.

The possibility that I might have to ditch had been in the back of my mind ever since I began losing altitude, but I never thought it would come to that. It was obvious now that there was no way I could make it back to the airport, and, much as I dreaded it, I would have to take my chances with the ocean. While I preferred the water to flying into the side of a hill, the odds of surviving a ditching at night in a stormy ocean weren't much better.

The clouds probably didn't extend down to the water, but as dark as it was, I still wouldn't be able to see the waves. Nor would I be able to judge my height over the water, so I would have no idea when to raise the nose of the airplane and flare out for the landing. Even if the airplane didn't break up on impact with the water or sink immediately, I would still have to crawl over the fuel tank and climb onto the wing to be ready for the search-and-rescue helicopter. Visibility over the water was less than half a mile, which would hinder rescue attempts. How long would I have to wait for the helicopter? No matter what the answer, it would be too long, because I would probably freeze to death in a matter of minutes.

In my sixteen years of ferrying airplanes, I had learned to live with the constant fear of dying alone in the ocean. All ferry pilots did. We might joke about it, but death was always there, lying in wait for just the right moment. For me—as I descended blindly into the black void—that moment was fast approaching.

The airplane descended through eighteen hundred feet. I still had about three minutes before I reached the hills and still enough time for something to happen. I didn't need much: a small tailwind to stretch my descent or an updraft to gain a little altitude—anything. St. Johns Airport was only a few miles from the ocean, so I could delay the decision to ditch until the last minute. Twelve hundred feet was the lowest I could descend and still have room to avoid the hills. If I didn't break out of the clouds by then, I would have to make the turn. I only hoped I could bring myself to do it.

At fifteen hundred feet, the airplane wasn't coming down as fast as it had been earlier, but not slow enough that I could reach the airport. If I had turned around two minutes sooner, I would have probably been able to make a safe landing. But I didn't, and now I would end up ditching so close to the airport I might be able to see it if not for the storm.

My gaze was constantly changing—from the windshield and the hope of spying anything other than the gray murkiness, to the airplane's altimeter whose needle showed thirteen hundred feet. The hills were only two hundred feet below me and closing in each second. The airplane dropped through twelve hundred fifty feet, too close to the minimum altitude to delay any longer. I began a turn toward the ocean, steeling myself for the ditching—when I broke out underneath the overcast.

The city of St. Johns spread out majestically before me, its lights cutting through the scattered wisps of clouds and bringing life to the deathly still darkness that had enveloped the airplane for what seemed an eternity. Whatever happened now, at least I could see where I was going. I felt as if I had been given a stay of execution. The airplane settled another hundred feet and then miraculously stopped descending. I was at one thousand feet, below the hilltops, but the invisible danger they once posed was gone as they passed harmlessly behind my wing. The airplane was barely flying fast enough to stay in the air, but at least it was flying. It maintained its hold on that precious one thousand feet of altitude as I turned back toward the airport and lined up with the twin rows of runway lights. With each second that brought me closer to the runway, I felt reborn. The strobe lights grew brighter and finally slipped beneath me as I closed the throttle and let the airplane settle onto the snow-covered runway.

It was two o'clock in the morning. I had been in the air less than forty-five minutes. That didn't seem anywhere near long enough to be so close to death, and then to be saved by a benevolent Trim God who, at the last moment, forgave my carelessness. Nor was this instance alone enough for me to reevaluate the dangers and the rewards of this addictive profession.

The banner days of ferrying small airplanes around the world had already crested, leaving fewer airplanes and more flights like this one—and more times that I would ask myself the same question. Was the adventure of this nomadic life worth it, when I knew that eventually the risks had to catch up to me and someday I might have to pay the ultimate price?

This is a difficult question to answer, if only for the fact that I was now safe on solid ground and able to ponder such thoughts. Once again I had cheated death of its prey, as I had done so many times in the past, since that day, which now seemed a lifetime ago, that I started out in this business.

2

The Call of Adventure

The problem was that I didn't know what I wanted to do for the rest of my life. Some days I thought about finding a stable job with a good retirement, getting married, having kids, and spending holidays at home with family and friends—everything exemplified by such dubious icons of cultural lore as Ozzie Nelson and Ward Cleaver. After all, this was the American dream that everybody who grew up during the dark ages of the Eisenhower administration was supposed to want. The next day I wanted to chuck everything, fly to Africa, and hitchhike down the Capetown to Cairo highway. I was constantly floundering between wanting a normal life and one of adventure and travel.

Adventure usually won out, because no matter how hard I tried to convince myself to settle down into a life of predictability and routine, I just couldn't get excited about it. While my contemporary baby boomers aspired to professional careers with the likes of IBM, AT&T, and General Motors, my interests were along the lines of signing on with a tramp steamer bound for the Far East.

Toward realizing the fulfillment of such a Twainian pursuit, I followed every lead and cockeyed scam that sounded the least bit interesting. There was the civil war in Biafra, which ended a few months before I was discharged from the air force. Then the war in Vietnam ended and curtailed any plans I had had about flying for Air America. When the Yom Kippur War broke out in 1973, I called the Israeli Embassy to volunteer, but was told they weren't taking foreign nationals. I even thought about joining the French Foreign Legion to throw my lot in with the other miscreants and nomads of the planet; I wasn't quite ready to make that kind of commitment, however.

Based on the direction my job campaign was taking, it probably sounds as if I was some gun-crazed, survivalist mercenary who got a kick out of blowing people up. But that was far from how I saw it. I wasn't looking for a war where you actually had to kill people. My quest was for the Hollywood type of war, where the fighting was always for a noble cause against an enemy of faceless strangers, and if people died, they did it cleanly with no suffering.

As the romantics might say, wars and revolutions were a young man's greatest adventure where all the good and bad of life can be experienced in an instant. That is what I was seeking and what I didn't think I would have trouble finding. After looking for two years, I was getting nowhere until I saw an advertisement in *Trade-A-Plane*:

> Pilot to fly solo throughout
> the world. Contact Globe Aero,
> Box 555, Lock Haven, PA 17745.

That's all it said. Nothing about what type of aircraft or if the cargo would be freight, passengers, drugs, or guns. Anything that vague had to be dangerous or illegal, or probably both. But that's what made it sound so interesting. I circled the ad, and four months later found myself in Lock Haven, Pennsylvania, being briefed by Walt Moody.

"Ferrying light airplanes all over the world," Moody, the owner of the small company, had said when he called two weeks earlier. He interviewed me over the phone and offered me the job. "I can put you right to work as soon as you get here."

That was all I needed to hear. I gave two weeks notice to the flight school I had been working for, packed a suitcase, and headed for Pennsylvania.

Once again I was a loner, starting a new job in a new city and hoping I wasn't making a mistake. Nothing I had seen thus far looked as if this town or this hangar might be the starting point for the adventure I craved for so long.

Lock Haven had a washed-in grime and a musty odor that attacked my senses as soon as I stepped out of the car. Only two years had passed since Hurricane Agnes tore through central Pennsylvania and flooded the town under ten feet of water. For nine days rain had pounded the area, knocking out telephone and electricity, and washing bodies out of cemeteries. Homes, buildings, streets, and bridges dissolved like sugar cubes or were tossed around like toys as torrents of rain caused swollen rivers and streams to overflow their banks, engulfing everything for surrounding miles.

Even now, in 1974, a year and a half after the flood, the small airport at the outskirts of town didn't look as if it had entirely recovered. Nor did it make me very optimistic. The hangar across the runway from the Piper Aircraft plant was a dilapidated structure of cinder blocks and corrugated-iron siding that looked as if it had been thrown together decades ago with discarded materials and then left abandoned. The building wasn't quite sagging, but that was merely a condition I figured would be altered drastically during the next strong windstorm.

There was no sign out front claiming that anyone might actually be occupying the structure, but there were several cars parked along the side belying any impression I might have had of desertion. It looked like countless other hangars on innumerable small, country airports where pilots eked out a living by giving flight instruction, sightseeing rides, and the infrequent charter.

Moody quickly put my fears to rest when he gave me a more thorough briefing. "We fly to just about everywhere: Europe, Australia, the Far East, Africa. So if you like to travel and don't mind sitting in an airplane fifteen to twenty hours at a time, this is the best job in aviation," he said.

I was still new enough to flying to think size equated to quality—the bigger the aircraft, the better the job. Airlines were obviously at the top of this social structure, while general aviation—small airplanes and anything that didn't fit into the first category—were the bottom feeders. It was the Siberia of aviation, where pilots were underpaid, overworked, and, in the structured hierarchy of this lofty profession, treated like poor country cousins. Moody attempted to shoot holes in this common tenet right away.

"Everybody wants to fly for the airlines. Big airplanes, fancy uniforms, and you can't beat the pay. But you don't do anything yourself anymore. They've got dispatchers and schedulers and ground crews to do everything. Ferrying small airplanes is different, because if you don't do it, it won't get done. You have to do your own flight planning, analyze the weather and decide on the best route. You have to take care of customs, hotels, airline reservations, and everything else yourself. It's one of the last jobs in aviation where the pilot is still kind of like a pioneer and is responsible for everything. And," he added, stressing the importance of this fact, "you don't have a lot of regulations and paperwork."

I had been flying for only six years but could see that aviation *was* getting more complex. Already the FAA regulations were several volumes thick and getting thicker every year. With more regulation came more paperwork. I didn't have any firsthand knowledge about airline flying, but I knew they had even more regulations governing them, combined with a similar ratio of paperwork and forms. There were forms for everything and in every color, shape, and size imaginable: maintenance release, flight plan, weight and balance, fuel order, dispatch release, and so on.

There's a saying in aviation that an airplane is ready to fly only when the amount of paperwork required for the flight is equal in weight to that of the aircraft. While many pilots accepted this burden as a necessary evil, I did not, and apparently neither did Moody.

Moody started flying in the early days of aviation when a pilot was in total command of his aircraft and one's fortunes depended almost solely on one's ability, luck, and cunning. He flew fifty-two combat missions as a bomber pilot with the U.S. Army Air Force in World War Two and saw action at the Battle of the Bulge. After the war he drifted in and out of a number of flying jobs: an airline carrying passengers and cargo from Miami to Rome, Egypt, and Australia; an aquarium hauling tropical fish from Latin America to the United States; and finally, to ferrying small airplanes overseas for Max Conrad, the famous long-distance pilot.

It didn't matter to Moody whether his passengers had gills and lived in water or were two-legged and lived in cities. All he wanted to do was fly airplanes—big four-engine behemoths that soared through the thin, nether regions of the upper atmosphere or small puddle-jumpers that barely had enough power to get in the air without clipping the top branches of tall trees at the end of the runway. Moody liked the work so much that when Max Conrad went out of business after three years, he started his own company.

"You couldn't have picked a better time to get into ferrying," he said. "Business was pretty slow when I started—a few old warbirds and small airplanes to Europe and the Pacific. But then, around 1965, general aviation started coming to life. Piper, Beech, and Cessna came out with new models, stepped up production, and set up distributor networks overseas. It wasn't long before I had to rent this hangar and hire pilots."

After logging more than five million miles and circling the globe hundreds of times, the miles and the years began catching up with Moody. He was in his early fifties but looked several years older—too old to sit in a small, cramped cockpit for twenty hours at a stretch, getting tossed around in thunderstorms and worrying about having to ditch. He had quit flying a year earlier and now manages the business and assigns the pilots to various trips.

At present, there were about a dozen pilots, whose location he kept track of by means of a small blackboard on the wall beside his desk. Stenciled along the left-hand border were the names of about twenty countries—Australia, Austria, Belgium, France, Germany, and so on—and to the right, on strips of magnetic tape, were the names of the pilots I would soon be flying with. On one strip of tape near the bottom of the board opposite the line marked DOMESTIC was added a new name—mine. I imagined the exotic cities I would soon be flying to and the adventures I would have as that strip of tape moved from one country to the next.

For now, though, I was ferrying airplanes from the second Piper factory in Vero Beach, Florida, to Globe Aero's hangar in Lock Haven, where they were fitted with long-range radios and fuel tanks. It wasn't exciting work but it was easy—when the weather was good. When it wasn't, or at night when I couldn't see the ground, the only way to get into Lock Haven was to fly to Williamsport, twenty-five miles to the east, find the Susquehanna River, and descend low enough so I could keep it in sight while following its curving path to the airport. That was the important part, because ten miles either side of the river the ground sloped upward to three thousand feet to form two ranges of the Allegheny Mountains.

Flying domestic didn't pay much, but it gave Moody a chance to see if I knew how to fly, while he waited for two compatible airplanes going to the same distributor so he could send me on my first trip with an experienced pilot. Globe Aero was delivering between thirty and forty aircraft each month, so I didn't think I would have to wait long. Until that time, I flew the short, domestic trips and tried to get to know the other pilots when they wandered into town.

There was John Snidow, on furlough from Seaboard World Airlines, Ernie Kuney, unyielding and barely tolerant of having to deal with the idiots all around him, Lev Flournoy, the lady's man, Jim Bennett, the pessimist, Jack Seaman, Fred Weir, Ernie Wheeland, Nils Mantzoros, Bill De Long, and George Rice, a diminutive pilot for whom the gods seemed to harbor a personal vendetta.

With the other pilots flying around the world, finding someone in town long enough to exchange more than a quick hello was a rare occurrence that happened when they were between trips and came into the hangar to turn in their gear and find out where they were going next. While waiting to see Moody, they hung out in the breakroom, griping about airplanes and weather, analyzing the latest emergency, or swapping tales that sounded as if they should begin with something like, "There I was, at ten thousand feet, when . . ."

Any story even remotely prefaced with such a cliffhanger invariably went on to describe in gripping detail some assured disaster that, through plain dumb luck, was only narrowly averted. While most of Globe Aero's pilots had at least one such white-knuckled tale tucked away in their bag of memories, they tended to downplay anything they might have done to save the situation. Many of our customers, however, could claim no such allusions to modesty. They dropped in from time to time and entertained us with tales of heroism and flying skills that suggested the pilot could trace his genealogy directly to John Wayne.

One of those pilots was Ernst Krupp, top salesman for Henschel Flugzuege Werke, the Piper distributor in Kassel, Germany, and one of Globe

Aero's biggest customers. Krupp was an ex-Luftwaffe pilot who often stopped in for some serious hangar flying whenever he visited Lock Haven. Although the conversation might start out about ferrying, it usually ended with Krupp telling about his exploits during World War Two when he was Admiral Karl Doenitz's personal pilot. For us, this was a novel perspective of the war, since Krupp, as he never hesitated to point out, "vuz on der losing side."

Moody never took part in these get-togethers and seldom did anyone ever hear him talk about his wartime experience, except for the one time when Krupp, attempting to goad his contemporary into matching exploits, asked, "Und vhere vere you during der voor, Valt?"

Most of the time Moody was serious. He seldom joked; and a smile, even a brief one, was something that appeared on the rarest of occasions. This was one of those occasions. The beginning of a grin turned up the corners of his mouth as he said dryly, "I was at twenty thousand feet bombing the shit out of Germany."

The trouble with Moody was that you never knew if he was in a good mood or a bad mood. Walk into his office one day and he might give you anything you ask for; walk in another day and he would chew your head off for the smallest infraction, especially if he learned that you met with a customer and weren't wearing a necktie.

Moody felt the silk scarf and leather jacket days of flying were over. He wanted his pilots to project a more professional image. Though the mandatory necktie could be discarded when not in the company of a customer, at all other times when in any way connected to Globe Aero pilots were expected to be in proper attire. Even wearing jeans around the hangar, which one pilot had made the mistake of doing, was considered an offense tantamount to mutiny.

"I worked hard to build this business," he told the pilot, "so if you want to stay working for Globe Aero, don't ever come into the office dressed like that again."

So I didn't wear jeans. On a few occasions I even neglected the necktie, since I wasn't meeting with real customers, only the manufacturers. I had been shuffling airplanes up and down the east coast for a little over a week when Moody called me into his office, presumably, I thought, to remind me of his dress code.

"We've got a B-58 going to Bankstown. Phil Waldman will be leading you in a Navajo, and Fred Weir will be in an Aero Commander, so you'll have plenty of company. Phil's got a lot of time in the Pacific, so listen to him and you shouldn't have any problems."

I was surprised for two reasons: first, because I didn't expect Moody to send me overseas for another couple of weeks, although I didn't have any idea where Bankstown might actually be; and second, because the only B-58 I knew of was a four-engine air force bomber. I found out later that morning that Bankstown was in Australia just outside of Sydney, and the B-58 I would be flying was not an air force bomber but a small twin-engine Beech Baron. It was brand new, white with pale blue accenting, and looked powerful for its size. Beech is the Mercedes of light airplanes, a solid, well-built machine so devoid of defects that I couldn't figure out why a piece of two-by-four was dangling from the tail.

"That's so it doesn't sit on its ass when it's full of fuel," John Probst, Globe Aero's hefty chief mechanic, told me.

"How much fuel can it carry?" I asked, a little suspiciously.

"Enough to go from Honolulu to Pago Pago without you having to go swimming. But here," he said, while opening the aft cargo door, "take a look for yourself."

Just about all I could see were tanks—two large aluminum fuel tanks that went from the floor to a foot beneath the ceiling and from just behind the pilot's seat to as far aft as the baggage compartment. What little space that was left was crammed with folded-up passenger seats, moldings, and other loose hardware. With so much weight in the back end, the airplane's center of gravity moved aft of the main wheels, causing the tail—when there was nothing to stop it—to slam down on the ground. The two-by-four, which was supposed to be attached whenever the airplane was on the ground, prevented this from happening. There was no such safeguard, however, for when the airplane was flying, because airspeed would be sufficient to counteract the imbalance. While it sounded reasonable, I wasn't yet convinced as to the validity of this aerodynamic theory.

Further inspection did nothing to allay my apprehension. Up front, where the copilot's seat used to be, was a jumble of fuel valves and clear plastic hoses that connected the internal tanks with the airplane's fuel system all mixed together with electrical wires strewn about the metal floor like so much leftover spaghetti. Aside from a valve or welded seam springing a leak and flooding the cockpit with fuel, I also found myself wondering about the likelihood of combustion.

"What if one of these electrical lines starts sparking?" I asked, but already had an idea what Probst's answer would be. Undoubtedly that concern had been posed to him many times over the years, because he didn't wait for me to finish the question. He tried to keep a straight face, but the twinkle in his eyes gave him away when he said, "You better run like hell."

His attempt at humor told me what I suspected—there *was* no answer. Since for the next ten or twelve days I would be flying over the ocean, with water stretching as far as the eye could see, if the airplane *did* catch on fire and I *was* able to get outside, there wasn't much I could do. I had no parachute, no survival equipment, and only a small two-man life raft. This was all academic, because I probably wouldn't be able to climb over the tank to get to the door in the first place. What I had thought was a great adventure just yesterday didn't look so good now that I saw close-up what could only be described as a flying bomb.

3

Earning My Wings . . . across the Pacific

Apart from requiring more speed for takeoff and being a little squirrelly on climbout, the Beech Baron was stable enough that after a while I could go several hours without even thinking about fuel or in-flight fire. The tanks and fittings withstood a few rough landings and turbulence over the Rockies without leaking so much as a drop. I was still a long way from being comfortable in the airplane, but by the time I reached the west coast and the fuel tanks had held together just as Probst assured me they would, I figured with any luck they should last until I reached Australia.

Phil Waldman and Fred Weir were waiting for me when I taxied onto the general aviation ramp at San Francisco International Airport. Weir was a big, lanky cowboy from Cody, Wyoming, who gushed with even more enthusiasm than I did at the idea of ferrying airplanes over the oceans. As for Waldman, he was like a mother hen, wanting to know how the airplane flew and if I felt comfortable in it. When my answers satisfied him, he started in on the groceries. Did I get this? Did I get that? Again I gave the answers he wanted to hear, until the next one.

"Did you get a milk carton?"

I vaguely remembered him telling me something about a milk carton when we were back in Lock Haven, but now I couldn't remember what. Waldman, however, placed a great deal of importance on this one item over all others and frowned when I admitted that I didn't have one.

"It's going to take us about fourteen hours to get to Honolulu and probably longer to Pago," he said, with an edge of sarcasm to his voice. "Did you think you might want to take a leak sometime?"

Though leaks had been uppermost in my mind for the past few days, they were of a decidedly different chemistry. Embarrassed over this obvious omission, the only thing I could think to say was, "Gee, I guess I forgot." It was an excuse that did nothing to enhance my credibility.

In addition to training me on this trip, Waldman was also Moody's assistant. He was one of the first pilots that Moody had taken on—a young kid without much flying experience who wasn't one to let thinking get in the way of acting. For most people this kind of attitude often led to misfortune, but in Waldman's case, it was what got him into flying in the first place. In the mid-1960s he hired himself out as an engineer with a diamond-prospecting operation in British Guyana only to learn upon arriving at the company's office in Georgetown that all the equipment had been stolen. Rather than go back home, he stayed and worked in the jungle with the rest of the miners.

Back then Waldman was a gullible kid who would go anywhere and do anything and love every minute of it. Dredging alligator-infested streams for precious stones and living off alligator tails in the jungle was an adventure beyond his wildest dreams. He made friends with a Canadian pilot who worked for the company and began helping out with the Canadian's string of businesses, which at times included a nightclub, a dance studio, a school for training circus acrobats, and a flight school. In exchange, the Canadian taught Waldman how to fly. With all the gems he had dredged up, Waldman bought an airplane and began flying miners in and out of the jungle.

When he returned home to Boston two years later, Waldman ferried a few crop dusters to Guatemala before landing a job with Moody and his new company. He was soon flying around the world having the time of his life when adulthood sneaked up on him. One day he was a carefree bachelor flying up and down the coast buzzing sunbathers, and the next he was married and being groomed by Moody to someday take over the business.

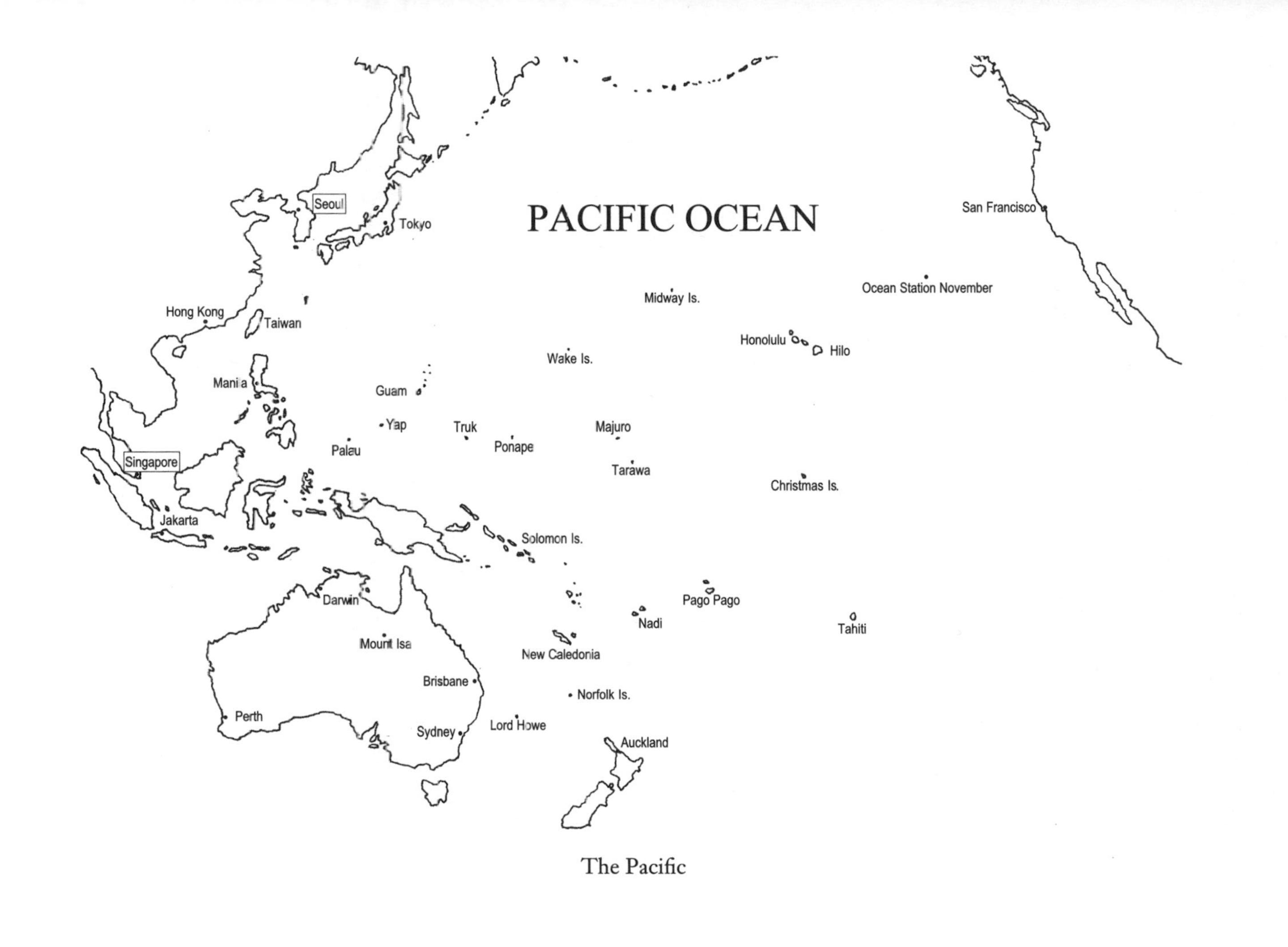
PACIFIC OCEAN
Seoul
Tokyo
Hong Kong
Taiwan
Manila
Singapore
Jakarta
Guam
Yap
Palau
Truk
Ponape
Wake Is.
Midway Is.
Majuro
Tarawa
Honolulu
Hilo
Ocean Station November
San Francisco
Christmas Is.
Solomon Is.
Darwin
Mount Isa
Brisbane
Perth
Sydney
Lord Howe
Norfolk Is.
New Caledonia
Nadi
Pago Pago
Tahiti
Auckland

The Pacific

The relationship of mentor and protégé was one characterized by tolerance and the ability to get on each other's nerves. I could always tell when they were having an argument or when Waldman just got yelled at by Moody, because he would take it out on the first person to cross his path. His tone was brisk and to the point, giving the impression that ferrying small airplanes was serious business and not to be taken lightly. He maintained that persona when he briefed me on what to expect while flying over the two thousand miles of ocean between San Francisco and Honolulu—after I had procured the all-important milk carton, of course.

"There's all the usual problems, like engine failures, in-flight fires, and a lot of smaller things that can snowball into a real emergency," Waldman said, which made me wonder just how usual these problems were. "But the biggest thing you have to worry about is navigation."

In the mid-1970s small aircraft were equipped with two basic types of navigation radios: the variable omni range (VOR) and the automatic direction finder (ADF). The VOR emitted stable, line-of-sight signals three hundred sixty degrees from the station. Its only drawback—especially when flying over the ocean—was that the signal couldn't travel over the horizon and was useless beyond one or two hundred miles. The ADF received commercial broadcast stations and non-directional beacons (NDB). While these signals did travel over the horizon and, depending on the strength of the transmitter, could have a useful range of several hundred miles, they weren't anywhere near as stable as the VOR.

"All this works great over land, with VORs and NDBs all over the place, and if the ocean was just a few hundred miles wide, ferrying airplanes would be a snap. But with thousands of miles of open water to fly over to reach places like Honolulu, we navigate the same way Lindbergh did to Paris—by dead reckoning."

Dead reckoning, or D/R as it's more commonly known, is simply taking the forecast winds for the altitude and route we planned to fly, calculate how the wind would affect the movement of the aircraft over a plotted track on the ground, and correct for drift. Like most pilots, I understood the inadequacy of forecasting wind direction and velocity twelve to eighteen hours into the future, and I was skeptical of our navigating like this over long distances with any degree of accuracy.

"With all the airplanes we've ferried in the Pacific and the Atlantic," Waldman assured me, "we've never been off course more than a hundred miles, and usually it's a lot less."

On these reassuring words we broke off the training to get some rest. Six hours later, at just after midnight, we met downstairs in the restaurant. Over several cups of coffee, Waldman brought up a subject I had tried not to think about for the last few days.

"This is your first crossing, and you don't know what's going to happen," Waldman began. "So if you start having second thoughts about continuing or if you have to return for any reason, go ahead and turn back to San Francisco. But Fred and I are going to keep going to Honolulu."

Apparently this was not new to Globe Aero. Several times in the past pilots had returned to San Francisco after reassessing their faith in things mechanical. More often than not, they just abandoned the airplane, leaving Globe Aero to bear the costs of straightening out the logistical mess.

"And usually," Waldman added, "it's not until the pilot's been over the ocean for several hours and can see nothing but water in any direction that he has this overwhelming urge to return to the safety of solid earth."

I didn't think this would happen to me, but then I was still on the ground and my confidence hadn't yet begun to wane. Or maybe it had a little. As we taxied to the runway to leave behind the safe familiarity of the North American continent for the vast expanse of the Pacific Ocean, I began questioning myself. Did I forget anything? Did Waldman forget to tell me anything? Everything was happening too fast.

The control tower cleared us for takeoff. I was right behind Waldman and Weir rolling down the runway and picking up speed. The next minute we were climbing out through a thin overcast in loose formation, the aircraft struggling under the weight of the extra fuel. I broke out on top of the clouds at thirty-one hundred feet and spotted the blinking red, green, and white navigation lights of Waldman's airplane a quarter of a mile ahead and to my right.

Oakland Center cleared us direct to the Farallon Islands, a low frequency radio beacon thirty miles west of San Francisco and the last contact we would have with land until the Hawaiian Islands. We leveled off at six thousand feet, and I settled in for the long ride of nothing but ocean and stars. I felt secure in my cocoon, protected from the elements of wind and water. All seemed so placid that it was difficult to imagine the ferocity of the ocean a mere mile below me. As I steered toward an unseen speck of land two thousand miles away, the impact of my folly began to sink in.

I thought about how the successful completion of the flight rested on the accuracy of the magnetic compass—probably the cheapest instrument in the airplane—and the mechanical integrity of the twin 285-hp engines. These engines, while finely tuned instruments of power on the outside, were a broiling inferno of timed explosions on the inside. Pistons were slamming against cylinder walls with the force of a jackhammer, and high-octane aviation gas was coursing through the fuel lines at a rate of seventeen gallons every hour.

If one of those engines were to quit, how long would the airplane stay in the air? Under normal conditions I knew I could maintain altitude on one engine, although the airplane would fly much slower. But these were far from normal conditions. With the extra fuel, the airplane was close to a thousand pounds overweight and would drag me down to a cold, black ocean in a matter of minutes.

Now was the time Waldman had warned me about. The lights of San Francisco had disappeared behind a black and formless horizon, and without their earthly reference we seemed suspended in space, unmoving against the night sky. The San Francisco VOR faded out at one hundred twenty miles, leaving us with no means of navigation except the series of D/R headings we hoped would bring us within range of the radio beacon on Ocean Station November, a U.S. Coast Guard ship permanently anchored halfway between San Francisco and Honolulu. I tapped the compass to reassure myself it was still working.

After several hours the night sky still hadn't changed. There was no moon, and, besides the flickering lights of my companions' aircraft, the only interruption from the deathly stillness were the infrequent strobe lights of an airliner passing six miles above us. The steady sound of the twin Continental engines droned on. Every now and then the radio came to life, as Waldman or Weir called in to compare navigation. Hearing another voice and being close enough to see the position lights of either aircraft helped to dispel some of the loneliness. And knowing that they were in similar craft only a few hundred yards away, even though they couldn't help if I got into trouble, was immensely comforting.

"Hello. Hello. You still awake?" Waldman's voice came over the speaker.

"Yeah. I've still got you at two o'clock," I said into the microphone. "Everything seems okay, but I'm ready for some daylight. How much longer do you think we have?"

"If you look behind you, you'll see the sky about to spit morning."

I looked outside toward the rear of the airplane and followed a black sky as it softened to pale blue and, finally, a faint hint of red just beginning to spill over the cloudless horizon. Shortly after we flew over Ocean Station November, I began receiving a broadcast station on Honolulu. It was mostly static, and the needle on the instrument was swinging crazily back and forth. I called Waldman to see if he was having better luck.

"I've got something, but the signal's still too weak," said Waldman. "This is the 'sunrise effect' I was telling you about that plays hell with medium- and high-frequency signals this time of day. Are you getting anything, Fred?"

"Mine is making me dizzy, it's bouncing around so much," Weir said. "I'm getting the same thing from the broadcast station on the big island, too."

"That's what I figured. They should settle down in another hour, once we get a little closer."

One hour later, just as Waldman predicted, the needle ceased its scatter-brained dance and pointed five degrees to the left. Somehow, against what would seem to be incredible odds, we were just about where we were supposed to be. We held our heading until we began receiving the Honolulu VOR, whose more accurate signal indicated we were only thirty miles north of course.

"They get a little nervous when we're off track," Waldman said. "Thirty miles isn't much, but no miles is better."

After maneuvering to the course our flight plan promised we would be on, Waldman called in our position. Honolulu Approach Control vectored us into a pattern with the rest of the inbound traffic and, as our turn came up, cleared us for the formation landing. We had left San Francisco exactly thirteen hours earlier.

During that time, I had gone from anticipation over the crossing, to worrying about the dangers, to relief when we sighted the island and began the approach, and, finally, to feelings of triumph, amazement, and gratitude for our success. I was both physically and mentally exhausted, and all I wanted was to find a hotel and sleep for twenty hours. Waldman, however, had other ideas and called for a fuel truck as soon as we taxied to the parking apron and shut down the engines.

"Why don't we refuel tomorrow?" Weir suggested, as he climbed out of the airplane and stretched his long frame to get the blood circulating again. "We have to come out to the airport anyway to export the airplanes and check the weather."

"With three airplanes it'll probably take them a while to refuel us," Waldman said. "I don't know about you guys, but I don't want to spend all day hanging around the airport. We're here now, so we might as well get it out of the way while we're still wide awake."

I didn't know who Waldman was talking about, because I wasn't wide awake. I was dead tired. I was stiff from sitting in one position and had airplane breath from breathing the same stale cabin air for all those hours. Granted, we would have a full day tomorrow getting ready for the flight to Pago Pago and then trying to sleep so we could leave that night, so Waldman's thinking did make sense. But I didn't see why we had to stick to a night schedule for every flight.

"The best time to arrive at a small island in the middle of the ocean is around midday, before it gets too hot and the clouds develop into full-grown thunderstorms," Waldman explained. "Because of the long legs, the only way we can get to wherever we're going by late morning or early afternoon is to take off around midnight."

"Why don't we just come out here a few hours before takeoff and refuel then?" I asked.

"Because when we come out here tomorrow night, it's a good possibility we'll be the only ones here," he said. "And it gets worse once we leave Honolulu. Pago and Nadi aren't too bad, but just about every other island shuts down as soon as it gets dark. Even if you can find someone at the airport late at night, I guarantee they won't be too ambitious about refueling an airplane. After a few trips you'll find it's best to put the airplane to bed first, right after you land. Then you can take off whenever you want with a minimum of hassles."

Waldman was repeatedly doling out little tips like this that would come in handy on later trips. I remembered them only because he stressed their importance so absolutely. He was still all business when it came to training, but the serious facade he had started the trip with was fading the farther we got from Lock Haven. Now, back in his old playground, he was joking and kidding as if he were just one of the guys. If this kept up I would have to reevaluate my original impression.

Aeromarine's sexy, blond receptionist looked up and smiled as we entered the trailer to pay for the fuel. "Well hellooo, Phil," she cooed. "We were wondering if you were ever going to come back and visit us. And now three ferry pilots all at one time! Gee, to what do we owe this honor?"

"Why, we came out to see you, doll," Waldman flirted.

"Aw, you're just saying that, Phil. I'll bet you really want to see Floyd."

"Where is the big kahuna, anyway?"

"Is that Phil Waldman I hear?" said a voice from the office behind the receptionist's desk.

"I don't know anyone else who comes here for fuel," Waldman shot back.

"So the prodigal son returns," said Floyd Goodyear, pilot, manager of Aeromarine, and all-around man about town, as he came out from his office. He was sufficiently lacking in girth or height to be mistaken for a big kahuna, but he seemed to enjoy the title. "Kuney and Seaman were here just last week, and now I see Globe Aero's got another pilot," he said, shaking Waldman's hand as if they were long-lost buddies. "Business must be pretty good."

"We're doing okay. We're buying a lot of fuel from Aeromarine. Speaking of which, when are you gonna give us a break on the price?"

"I'm working on it, Phil. It won't be long now."

"What, is that what they said when you were circumcised? Come on, Floyd. I think we ought to talk about a volume discount," Waldman said.

"Phil, I'm hurt! I'd like to give you a discount. Honest. But I got this girl I've been going out with who wants me to run off to Maui with her. You gotta see her—she's an absolute knockout. And all she wants to do. . . ." Goodyear rambled on, a roguish grin spreading across his tanned face. He seemed to grow several inches, and his eyes began to twinkle at the memory of blissful groping between the sheets.

As Waldman had told me during the crossing, it didn't matter what Goodyear was talking about, sooner or later the topic got around to the finer attributes of a particular female he had recently had the good fortune to spend the night with. From anyone else these exploits would probably be relegated to the realm of wishful braggadocio, but Goodyear—handsome, suave, silver-haired, with a silver tongue to match—told the story with such sincerity, that whether real or of his own creation, you wanted to believe him.

One of the myths—which he did nothing to dispel—was about the affair he had with a Pan American stewardess based in Honolulu who lived in a plush apartment, drove an expensive Jaguar, and had loads of money that she parted with as freely as her other favors. All of which would appear to be beyond the means of even the most senior stewardess. But if Goodyear was curious about her extravagant lifestyle, he didn't question it. After all, she was spending much of this money on him, and it was he who drove the Jaguar when she was flying around the world.

Life for Goodyear was indeed good, and might have stayed that way if it hadn't been for "The Thrilla in Manila." He was watching the Muhammad Ali–Joe Frazier boxing match on closed-circuit TV when the camera zoomed in for a close-up of Ferdinand Marcos enjoying the fight from the presidential box. Sitting beside him was none other than Goodyear's worldly girlfriend.

Besides Goodyear, about a million or so other people were also watching the fight, one of whom was Imelda Marcos, the president's wife. And her reaction to her husband's extra-marital dalliance was decidedly different from Goodyear's.

Loyalties in the Philippines were divided. There were those loyal to the government, the army, the political parties, and the rebel groups. There were also those who were loyal to the president and others to the first lady. Imelda ordered that the girl be taken care of, but before her loyal minions could carry out their sinister mission, another group of generals loyal to the president went to the girl's hotel room in Manila and sneaked her out of the country.

To Goodyear, the thought of dictators, first ladies, and generals hatching secret plots and skulking around palaces was something only Rocky and Bullwinkle at their most ludicrous could choreograph. All this palace intrigue did wonders for his ego and made sporting around Honolulu in Marcos's Jaguar that much more enjoyable. As with most myths, this one was a

that will make the outcome as painless as possible. If you do get lost and end up having to ditch, your chances of surviving are a lot better if you can see the waves before you plow into them."

With no ocean station between Honolulu and Pago, it made for a long and worrisome flight. Finding a tiny speck of land after flying over two thousand miles was in itself something to boast about when trading stories with other pilots. But when we began receiving the Pago Pago broadcast station and it showed us to be precisely on course and on time, I figured it an event of majestic proportions.

There were a few harmless puffs of cumulus hovering over the island and a larger, darker one building over the Rainmaker Mountain and obscuring the peak. We easily avoided it as we descended over hilly terrain covered thick with brush and jungle. A short distance inland banana-clad hills rose into the heartland, which was pitted with green-covered craters of inactive volcanoes. Turning onto final approach, I saw the scorched earth and plowed hillside where a Pan Am Boeing 707 crashed several months ago while attempting to land during a thunderstorm. It was a reminder that made me glad Waldman was so insistent we arrive as early as we did.

After refueling the aircraft, we trooped into the airport restaurant to grab a quick breakfast. Or maybe it was lunch. With the long night flight and having flown through so many time zones in the last few days, my stomach had no idea which particular meal would be proper, it just demanded food.

Outside, a Polynesian Airlines Twin Otter landed, and a mob of colorful Samoans, mostly huge women carrying large bundles and tiny babies, swarmed out to greet relatives and friends. Just as this happy reunion was getting underway it began raining. Pouring would be a more accurate description, for literally out of the blue a dark storm cloud drifted overhead and loosed a barrage of water onto the tiny airstrip. In a short time there were great puddles of water everywhere, and anyone foolish enough to venture outside soon came rushing back in, bringing with them a blast of rain as the gale winds ushered them through the doorway.

Thirty minutes later the sun was shining, and people continued on with their routines as if this momentary deluge was just a commonplace inconvenience, which, to the islanders, it was.

We were in the air shortly after finishing our meal and heading for Nadi in the Fiji Islands, a short, seven-hundred-mile leg that we made in four hours. It was just getting dark when we landed at Nadi. And though we had been up nearly twenty-four hours with only four hours sleep in Hilo, we still kept to the same grueling pace of refueling and tending to the aircraft before addressing our own fatigue.

Globe Aero's policy was to arrive at the customer's hangar during working hours, which would normally allow us to get a decent night's sleep. But since Waldman had an additional five hundred miles to fly to Melbourne, we moved our departure time back by four hours so we could fly as far as Sydney together. Once again, we got up when most people were going to sleep and began our workday while the sun was over another hemisphere and sunrise was still several hours away. By now I was getting used to this nocturnal schedule and accepted it with a resigned shrug.

It was our last leg—a seventeen-hundred-mile run to Sydney. After the seven thousand miles we had just flown, it seemed almost too easy.

"Don't let it fool you," Waldman said. "It may be summer down here, but this part of the Pacific can be as unpredictable as the North Atlantic. I know you haven't been across the Atlantic yet, but take my word for it, you can run into cold fronts, icing, turbulence, and hail any time of the year."

Weir confirmed Waldman's warning. "Sydney's pretty far south of the equator; almost as far south as New York is north. I don't think I've ever been down here when I didn't have to fly through a front somewhere between Nadi and Sydney."

"Fred's right. And you can't depend on what the weather office tells you. They try, but they just don't get that much information, and they usually blow it. Always plan for at least a thirty-knot headwind and even then take plenty of reserve fuel."

We were halfway to Sydney when the thin, scattered layer of cirrus above us began to fill and gradually slope downward to merge with a lumpy gray line on the horizon. Thirty minutes later it formed a solid wall of clouds that looked impenetrable. With no way of flying around it, we separated and flew at altitudes two thousand feet apart and slowed the aircraft to lessen the risk of structural damage. Then we hit the turbulence.

I might as well have just been along for the ride, because this roller coaster sent the airplane hurtling upward at two thousand feet per minute. I attempted to stop the crazy ascent by closing the throttle and pushing the nose down. No sooner did I take this action, than a downdraft hurled the airplane in the opposite direction. I hastily tried to correct the changes I had just made. We bounced around like this for an hour before the last cloud spit us out into a clear blue sky that carried us the rest of the way to Sydney.

Two reassuring squeals of the main tires and the thump of the nose wheel touching the runway were followed by the control tower directing me to the international parking ramp. Weir had landed just ahead of me and was waiting by his airplane. We made a quick dash into the terminal to clear customs and then jumped back into the aircraft for the short, five-minute hop over to Bankstown.

A ferry pilot's time, I was learning, was a fleeting commodity. Either we had hours weighing heavily on us in airplanes plodding across the oceans or days on the ground waiting for better weather. Then there were times like this, when we rushed to get things done and every second counted.

Before we left Nadi, Waldman told us there were several more airplanes in Lock Haven and urged us to hurry back and make a quick turnaround. It seemed a waste to come all this way and then get on an airline to fly right back to the States without doing any sightseeing, but from what I could gather, that was what the ferry business was like. Waldman, Weir, and the other pilots all seemed to hurry back to Lock Haven, take a few days off, and then were gone again.

Being the new guy, I didn't think I should deviate from this established practice so soon, nor did I want to. Six months ago I wouldn't have believed a job like this existed. Yet here I was, given an airplane and all I had to do was fly to Sydney, Australia, or maybe Nykoping, Toussus Le Noble, or any of the other cities nonchalantly mentioned in a ferry pilot's conversation. In all honesty, I couldn't wait to get back to Lock Haven to find out where I would be going next. Besides, I figured I would be coming back to Australia in the months to come and have plenty of opportunity to see the country. So for now I settled for a quick view of bustling, downtown Sydney through the taxicab windows.

As the Qantas Airlines Boeing 747 lifted off the runway and began a shallow turn to the northeast, I adjusted my seatback to as comfortable a position as possible. Since the last time I slept in a real bed, I had seen the sun go up and down twice. Now it looked as if I would witness that cycle two more times before I got a chance to renew that acquaintance. Then it's back into another airplane and do the whole thing all over again.

4

The Route to Singapore

The Arab oil embargo, which reached its peak in the winter of 1974, had disrupted transportation around the country, causing motorists to wait in long lines for a rationed amount of gasoline. The spot shortages also affected aviation in the Pacific. Many islands were either out of fuel entirely, or supplies were so sporadic that availability was often in question. This was something I didn't think about when Moody told me I was going to Singapore.

Of all the great cities in the fabled and exotic Far East, I had always thought of Singapore as the one place where you could find adventure around every corner. I imagined that Chinese warlords trafficked in opium, guns, and bounties of the flesh, American and foreign intelligence agents traded secrets across the tables in cafes and nightclubs, and seduction by exotic femme fatales was not only a possibility, but quite likely.

Who could think about fuel with characters like these roaming in and out of their thoughts? I couldn't, but Moody could. He told me that because there would be two airplanes, which would require double the amount of fuel, we would have to plan our route carefully.

"Bennett's taking the other airplane," he said. "He's been to Singapore and knows his way around the Pacific, so you shouldn't run into any problems."

"The closest I've been to Singapore is five thousand miles," Jim Bennett said, when he came into the hangar later that morning. "And I don't think anyone else with Globe has been there either, so we're going to be the guinea pigs."

"I guess the normal route after Honolulu would be either Marjuro or Tarawa, then Truk, Brunei, and Singapore," Bennett speculated, as he looked over the map. "But most of those places are out of fuel; I know Majuro is and so is Tarawa. Pago should be okay, but after that, who knows? We might be able to go to Honiara, but I don't know if they'll have any either. Australia's probably okay, but I've never been farther than Sydney, so I don't know what to expect."

What should have been an almost straight great-circle route from Lock Haven to Singapore now resembled a hyperbolic curve that dipped well into the Southern Hemisphere before turning northwest to Singapore. It added eleven hundred miles to the trip—more if we had trouble finding fuel. It seemed there was a lot more pioneering and uncertainty to this job than Moody had led me to believe.

"How long do you think it'll take to get there?" I asked.

Bennett thought about it before answering. "I don't know. I can't see us flying all that way without something going wrong. Just to be safe, I'm going to pack for three weeks."

Although this wasn't a training flight, I suspected Bennett was under instructions from Moody to see if I could navigate without someone leading me. I didn't mind this arrangement but apparently Bennett did, for he liked neither training pilots nor being a spy for Moody and was wary of letting his guard down. He used his words sparingly, as if small talk and idle chatter were better left to fools and blowhards, a subspecies he believed one should always try to avoid. He wore a pencil-thin mustache over his lip that added a sharp edge to his features, and his dark, riveting eyes bespoke a challenge to anyone who might waste his time with inanity.

During the first few days, Bennett's comments were confined to the technical aspects of our flight. He began to open up during our layover in San Francisco, but it wasn't until we reached Honolulu, and I had dispelled any fear he might have had about my making the next two weeks of his life miserable, that he let his guard down and we started to become friends.

Honolulu had seen little change in the weather since the last time I was here. The winds, blowing steady out of the southwest, had been rising and falling in velocity and were presently on the increase. The Hawaiian briefer called them *Kona* winds and, like my last trip, he wasn't optimistic about any improvement on the route to Pago Pago for another five or six days.

"I figured something like this would happen. We're not even halfway there and already we're running into delays," Bennett said. "I guess all we can do now is send a message to Honiara to see if they have any fuel. Because when we do leave here I'd like to have some idea where we're going after Pago."

It took two days to get a cable from Honiara advising us that they were out of fuel. Because the winds hadn't changed, we still had nowhere to go except back to Wakiki for more sightseeing. We ended up at Davy Jones's Locker, a small beach bar in the basement of the Reef Hotel, where our conversation eventually turned toward ferrying and the pros and cons of our vagabond profession.

"It's an ego trip," Bennett said. "Not many people have even heard about ferrying small airplanes overseas. Hell, there aren't many pilots that know anything about it. So when you tell someone you're flying to Pago Pago or Singapore, it probably sounds pretty exciting."

Of course it sounded exciting. These were places John Wayne and Humphrey Bogart used to hang out. Except I couldn't remember them ever worrying about fuel. "I'd be a lot more excited if these airplanes had another twenty gallons in them," I suggested.

"Now, now, we can't have any more talk like that," Bennett said, admonishingly. "Walt would probably tell you that back when he was ferrying airplanes they had the same fuel and it was enough then. Maybe you just need to be more efficient."

The conversation got around to the many problems we decided had Moody stymied. We had been at it for three days when Lev Flournoy and Bill De Long arrived. I had met them both in Lock Haven and could not have imagined a more unlikely pair of traveling companions. Flournoy was tall with an athletic build and looked like he belonged on the cover of a romance novel. While De Long, in corduroy jacket, sneakers, and ever-present pipe, reminded me more of an absent-minded college professor than a pilot who used to fly for Air Vietnam during the war in South East Asia.

Following the pattern Bennett and I had set, we continued our barhopping forays. Now, with the security of numbers, we ventured into areas of lesser repute. We began with Hotel Street, Honolulu's red-light district, where a menagerie of pimps and prostitutes, pickpockets, junkies, and other assorted denizens of the night prowled the streets.

"So this is how you guys spend your time," Flournoy joked. "Walt sent us to check up on you and make sure you weren't taking a vacation. When you gonna get moving, anyway?"

"I don't know," Bennett said. "When are the winds going to straighten out?"

"If you're going to worry about winds, you guys should be flying Cessnas," said Flournoy, seizing the opportunity to gloat over our misfortune in flying aircraft with such limited range. "Hell, we've got enough fuel in our airplanes to go nonstop to Nadi if we want."

"Uh . . . speak for yourself, Flournoy," De Long said, as he sucked on his pipe. "I'm not crazy about flying to Pago when we've got a tailwind; I know I'm not going with a headwind."

De Long was not one of Globe Aero's faster pilots. While it might take anyone else three or four days to fly a European trip, De Long was gone a week or longer. During that time, Moody rarely knew where he was and often sent another pilot on a trip, telling them "Be on the lookout for De Long. He left several days ago, and no one's heard from him since. See if you can light a fire under him to get him moving." De Long was the sole exception to the speed and efficiency credo Moody tried to instill in his pilots, taking so long to deliver an airplane that he became known as "Long De Long" or "Mach Minus," an inverse reference to the speed of sound.

"So you guys are really going to Pago, huh?" Flournoy said. "Sounds like a long way to get to Singapore."

"It gets worse," I said. "Since Honiara's out of av gas, it looks like we're going to have to go through Australia."

"But that'll take an extra three or four days," he said, pretending to be astonished. "I'd slit my throat if I had to fly that much longer."

“And Walt wants you to hurry up and get back to Lock Haven,” De Long chimed in. “Because he wants to send you right out again.”

“Walt always wants us to hurry back,” Bennett said. “What else is new?”

“I don’t know what you guys are complaining about. I’d be perfectly happy to stay right here in this bar,” Flournoy said, as he began eyeing an attractive young lady sitting by herself a few tables away. Flournoy fancied himself a lady’s man; tanned, handsome, and a smooth talker, he walked over to the girl and began plying her with drinks until she eventually succumbed to his charms. The rest of us, in admiration of his seemingly effortless conquest, watched enviously until he returned to our table with a look of disappointment as he sadly informed us that she was a he.

The next morning the winds had begun to weaken. In all, Bennett and I spent five days in Honolulu; now that we were finally able to continue, we were eager to begin. We tried to sleep in the afternoon, a chore that I have proven to be hopelessly incapable of. I did manage to drift off for a few hours before we took off late that night for Pago Pago—Bennett and I together in one flight and Flournoy and De Long thirty minutes behind us.

I was wide awake throughout the takeoff and climb, but after we leveled off and set course, there was little to do. The autopilot keeps the airplane straight and level much better than I can. Other than making a heading change and switching fuel tanks every few hours, there was nothing to do, and it was difficult to keep from falling asleep.

After several hours my eyes started getting blurry from staring at the instruments and, just as I was about to doze off, I would snap my head back and shake it from side to side. Ten minutes later I got cross-eyed and the instruments became fuzzy again. The steady vibration and monotonous drone of the engines were trying to lull me to sleep, but I kept fighting it. If I could just sit back and close my eyes for a few minutes, I thought, this sleepiness would go away. Just a few minutes. . . .

I was abruptly awakened by Bennett’s voice coming over the speaker. “Hey, are you there? Nadi’s calling. They want to know where you are.”

Still only half awake, I heard the words clearly, but their meaning eluded me. I continued rousing from my slumber, looking up, over the top of a strange panel with dimly lit dials. I stared outside the windshield, hoping to recognize something, when a feeling of dread overcame me. All I could see was a black sky dotted with twinkling white stars above me and dark, nebulous, moonlit shapes of a scattered cloud layer below me and the tiny red and green dots from Bennett's lonely airplane.

"I don't know where I am!" I screeched into the microphone.

No answer. The seconds ticked by, and I began to remember what was going on. I was in an airplane over the Pacific Ocean on my way to Pago Pago, and Nadi, the controlling facility whose sector I was flying in, wanted to know my position. I was relieved to learn that I was where I was supposed to be and thankful Bennett hadn't answered my delirious response right away, but had given me time to fully waken. At one time or another, he had probably gone through the same frightening experience. All I needed were those few minutes of sleep, because I was fine for the rest of the night, a little drowsy when the sun came up, but wide awake when we landed at Pago Pago.

Certainly the reliability of available fuel was reason enough to come to Pago Pago, but it wasn't the only one. That distinction was simply because Pago Pago was a cheap stop. Ferry pilots weren't paid a set salary, but were contracted for each trip and given a fixed amount of money to deliver the airplane. Out of this money, we paid all expenses: fuel, landing fees, hotels, meals, taxis, airline tickets, and anything else that might crop up.

Whatever money was left after completing the trip was the pilot's, which, if there were no major delays, was a reasonable sum. So, in theory, the amount that a pilot could earn was limited primarily by how efficient he was in delivering the airplane. The only problem was that the figures Moody used to calculate expenses were based on what he used to spend. Because Moody hadn't flown a trip in over a year, those figures were somewhat outdated.

In truth, we had to be downright miserly. We sought out the lowest airline fares. We doubled up, two and sometimes three to a hotel room. If our aircraft had enough range and we had good tailwinds, we overflew some of the usual stops to avoid paying additional landing fees and hotels.

The one thing we couldn't be stingy with was fuel. It was our biggest expense—at least half of everything else combined. Yet, the cost of fuel was rising fast, much faster than normal inflation. OPEC and the Arab oil embargo had seen to that. Fuel in the continental United States was still the cheapest you could find anywhere in the free world. But at a dollar fifty a gallon it was twice what it was before the embargo, and that was when it was available.

Even without the embargo, flying the extra distance to Pago Pago paid for itself because here, in the middle of the Pacific Ocean—where one might reasonably assume logistics increased the price of everything several-fold—fuel cost a ridiculously low fifty-two cents a gallon. This alone made Pago Pago more desirable than any other island, but an even greater incentive was that the meter on the fuel pump was so far out of calibration, it registered considerably less than what was actually pumped into the aircraft. I knew about the lazy meter from when I was here two weeks ago with Waldman and Weir, but figured it would be fixed as soon as someone noticed the discrepancy. Apparently no one had, because it was still broken.

"It's been screwed up as long as I've been coming here," Bennett told me, after he filled one of his airplane's two 100-gallon fuel tanks and was about to start on the second. The meter read fifty-six gallons. "Not bad, huh? This is why everybody at Globe comes to Pago."

Lowering his voice so the Samoan refueler couldn't hear, he further explained, "All you have to do is pump the fuel real slow so it's almost dribbling out of the nozzle. It takes a while to fuel the airplane, and sometimes they get grumpy if you take too long, but just tell them you can't pump any faster or you'll spill fuel all over the place."

With the temperature close to one hundred degrees Fahrenheit, I was already sweating. Staying outside even one minute longer than I had to was not something I particularly wanted to do. But after Bennett filled his second tank and the meter showed only forty-three gallons, I forgot all about the heat and started refueling my airplane—slowly.

Between both airplanes, I figured the meter had undercalculated the amount of fuel by almost three hundred gallons. Flournoy and De Long had just landed and were now refueling their aircraft. They did not appear to be in any great hurry.

"Doesn't anyone ever get suspicious when the numbers don't add up?" I asked Bennett, as we waited inside the air-conditioned snackbar, drinking ice-cold sodas and wondering how long we might be able to get away with this deception.

"I doubt it," said Cary Wade, the American mechanic for South Pacific Island Airways, as he joined us and offered his views on what was acceptable larceny. "Anyway, no one's ever said anything."

"What about when the tank's empty?" I asked. "What is it, a twenty thousand gallon tank? Doesn't it ever occur to anyone when the records show they only pumped—say ten thousand gallons—that something isn't right, and maybe, just maybe, they're losing money?"

"I guess you'd think so, ay," said Wade. "But apparently that's a system of inventory that hasn't found its way to this part of the world yet."

"So they'll just think the tank's leaking," said Bennett, who saw no need to question a good thing.

"Oh, they probably already know the meter's screwed up," Wade said. "It's just that nobody's in a big hurry to fix it. So they lose a bunch of money. Big deal! It'll all even itself out in the long run. Hell, they get more money from the United States than they know what to do with."

"See! What'd I tell you? This is where all my taxes are going," said Bennett. Since leaving Lock Haven, he had been expressing his indignation over the government's wasteful foreign-aid programs. "Now why in the world do we have to support a place like this? Just tell me that."

"Hey, we gotta keep the chief happy, ay," said Wade.

"As much money as we give him, he must be in pig heaven."

"And we appreciate it," said Wade, before turning to me. "So you see, you just have to think of all this free fuel as Pago's way of thanking you for being so generous."

Wade's philosophy toward fuel supplies and island life in general was simple: it was what it was and there was no use getting excited about it. He came to Pago Pago a year ago and apparently found the unhurried pace to his liking. His job maintaining the one small commuter plane that flew between the islands of American Samoa did not entail a lot of work, which left him plenty of time to hang around the airport and greet the ferry pilots when they landed. Extroverted by nature, he easily fit the unofficial role of ambassador of goodwill, often showing the pilots around the island and then taking them to the small bungalow the airline provided him with to barbecue steaks for everyone. The evening then followed a predictable routine of drinking and carousing at the Sadie Thompson Bar in the Rainmaker Hotel.

Sitting in the crowded bar packed with New Zealanders and Samoans, Wade offered his perspective of life on Pago Pago while the rest of us couldn't help becoming curious over the sight of several burly Samoan men dressed in sarongs and wearing a flower above their ear.

"The flower means they want to be courted—both guys and gals," Wade said. "Ya see, in Samoan culture, if a family has three or more sons, the third son is raised as a girl and wears dresses, puts on makeup, does women's chores, and goes into female occupations." Wade's observation may not have been as exhaustively researched as Charles Darwin's or Margaret Mead's, but he believed it and told it with such conviction that we believed it, too. While the effects of too many Australian beers might account for the few insensitive comments, good sense—owing to the size of the Samoans in question—kept us from making them too loudly.

The flight to Nadi the following morning was so close to the end of the trip for Flournoy and De Long that Flournoy felt it necessary to remind Bennett and me in case we had forgotten. "Only seventeen hundred miles and we're all done," he said, with a snide grin. "How much farther did you guys say you had to go?"

It wasn't only that Bennett and I still had one-fifth of the Earth's circumference to fly, but because of the delay in Honolulu, we didn't have a lot of time to do it. With reservations on Pan Am in five days and at the rate airlines were canceling flights, we didn't want to arrive late and have to wait several days for another one. Hoping to make up for lost time, we slept a few hours in Nadi, flew all night, and landed in Brisbane just after sunrise. It was the wet season in Australia when monsoons swept through the equatorial latitudes and it rained for days. Although Brisbane was clear, it wasn't expected to stay that way.

"You up to flying another twelve hours to Darwin?" Bennett asked.

"It doesn't look like we'll have much weather to fly through, except maybe some headwinds," I said, studying the charts. "I'm not too tired now; I guess we might as well get it over with."

Pumped up with several cups of coffee, we left Brisbane and headed west, into the barren landscape of Australia's outback. The next several hours took us over an endless sea of bleached-tan grass draped over red-brown earth, camouflaged with splotches of dingy olive scrub brushes, desolate under a scorching southern sun. We flew over long thin ribbons of road, isolated farmhouses, and dead earth—like some postapocalyptic wasteland that was lifeless except for the wandering herds of kangaroo or the occasional dingo.

In time, even those few signs of life disappeared as a layer of clouds began to form beneath us, scattered at first, but becoming denser the farther west we flew until the landscape was a solid blanket of white. The sun relentlessly bombarded the mirrorlike surface of the clouds, reflecting so much glare that I couldn't look outside without squinting and cursed my clumsiness for having broken my sunglasses in Brisbane.

The sky was darker to the west, signaling the approach of deteriorating weather that would probably hinder our chances of continuing on to Darwin today.

"It looks pretty grim up ahead," I said into the microphone.

"I was just thinking the same thing," Bennett answered. "Brisbane didn't say anything about it, though, so let's keep going for a while and see if we can get around it. Besides, there aren't a whole lot of places we can land."

We had been talking with Brisbane Radio on the high frequency set, but now, within range of Mount Isa, we called approach control on VHF and listened for reports about the storm. What we heard did not sound promising. One after another, airliners were calling in to request vectors around the storm's more severe cells—airliners that flew higher and faster than our small Pipers and were equipped with weather radar that showed the location of the cells. Bennett and I had neither the equipment nor the desire to fly blindly into the clouds and try to come out the opposite side in one piece. The storm was just too big—a cumulonimbus monstrosity that rose to thirty thousand feet. I was mesmerized by the sheer size of it and was tempted to see up close the dangers that lurked within: torrential rain, hail, and turbulence so severe it could twist my tiny airplane like a tin can. Humbly, I kept a safe distance.

"We've got a squall line of thunderstorms just to the west of Mount Isa moving east at ten knots," the controller warned us. "Radar is showing some heavy cells, but we're okay at Mount Isa right now if you want to land here."

The controller wasn't telling us as much as he was hinting at what he considered our safest course of action. That was good enough for Bennett and me. The sky was almost black when we told the controller we would land and wait out the storm. Just after we landed and finished tying down both aircraft, the rain started. It was gusting to thirty-five knots, and we gave up any thoughts of flying any farther tonight. Instead, we hitched a ride into town with a young kid who took us on a gut-wrenching, tire-squealing drive, dropping us off in front of the Mount Isa Hotel.

As the only place to get a room for the night, the hotel's owners apparently saw no reason to spend money repainting the weather-worn clapboard siding or make any attempt at sprucing up the place. It was faded and rundown, looking as if it had borne the brunt of too many hot Australian summers. Inside was just as dismal. An elderly desk clerk led us upstairs—pointing out the common bathroom we would share with the other guests on our floor—showed us to a room with two small cots, and disappeared back down the narrow stairway.

It was dark by the time we went outside to look for a restaurant. The streets glistened from the recent rain, reflecting red then green from the lone traffic light on the corner. It blinked needlessly, for the streets were deserted, and the few shops in the old brick buildings had long since closed for the night. We passed a Woolworth's Five and Dime, crossed the street, and headed toward a small, chrome-and-neon-lit diner. A car drove by with several joyriding teenagers who were looking for girls or mischief, or most likely both.

"Did we fly through a time warp or something?" I asked, thinking that everything seemed just a little too strange. "This place looks like a small town in the States back in the fifties."

"I think you're right. These guys are out cruising. Yeah. Where were you in sixty-two?"

Bennett's reference to the movie *American Graffiti* was right on target. Even the diner, with its Formica tabletops, vinyl-covered seats, and jukeboxes, might just as well have come straight out of America's rock and roll era.

The waitress eyed us suspiciously as we sat down at a table by the window. Did we look that different from everybody else? Did she already have us pegged as outsiders who might bring her grief at the end of a long and trying day? She gave us a few minutes to look over the menu before padding over to our table. This tired, frazzled woman looked as if she would take no gruff from us or any of the young punks hanging around.

"Can I help you?" she snapped like a drill sergeant, jotted down Bennett's order, and then turned to me with a terse, "And you?"

I read my order from the cardboard menu and handed it to her. But she didn't take it; she held her pencil firmly to her notepad and testily snapped, "Please!"

Thinking she didn't hear me or that maybe she couldn't understand my accent, I read from the menu again, pronouncing each word as clearly as I could. Then, like the first time, I tried to hand the menu back to her.

Still, she made no attempt to take it. The waitress didn't move, didn't flinch, and just glared at me, implacably. Louder now, and with an edge of weary authority, she barked, "*Please!*"

For the life of me, I had no idea what she wanted. Bennett was snickering, the waitress was impatiently tapping her foot, and I was beginning to feel as if I had committed some cultural blunder that was obvious to everyone but me. Again I looked at the menu, but before I could say anything, she scooped it up, put her pencil in her breast pocket, and, in a voice dripping with venom, issued a final, "Well! You don't have to be so rude!" She stormed off into the kitchen.

"Now you did it," Bennett said, chuckling audibly now that the waitress couldn't hear. "We'll probably never get any food."

"I must've missed something, because I still don't know what her problem was."

"Come on, you really don't know?"

"I don't have the slightest idea."

"She wanted you to say 'please.' "

" 'Please'? I thought I did."

"Apparently she didn't think so. And now you got her all pissed off. Good work. She probably thinks all Americans are jerks. So much for Mount Isa," he shrugged. "I guess we can't come back here again."

Mount Isa was so far removed from our normal routes that the chances were slim that we would ever find ourselves here in the future, and I couldn't see any reason to purposely come back. Bennett, however, felt differently.

The small town had made such a favorable impression on him that during the flight to Darwin, it was all he talked about. "A simple life in the outback where everyone's your mate." And after he had talked himself into it, "Yep, that's where I'm moving to when I retire."

Two hours on the ground in Darwin, and we were in the air again. "If we leave tonight, we can be in Singapore in the morning," Bennett had said, "have all day to look around the city, and still make the Pan Am flight day after tomorrow. Besides, the weather looks good."

The only problem was that it didn't stay that way. The southwest corner of the Pacific Ocean was a breeding ground for typhoons, hurricanes, and tropical depressions that popped up with little warning and plowed a path between Australia and Indonesia. We had reached the southern coast of Borneo when we ran into one of those storms, which, like the one at Mount Isa, did not show up on the weather charts.

We altered course to fly along the storm's southern rim and climbed to stay above the clouds. Seventeen thousand feet was as high as we could go, and even that was a struggle. Built to operate in the heavy atmosphere below ten thousand feet, the engines were gasping for air and showed a marked decrease in performance. It wouldn't be long before the rarified air started affecting us as well. I could still think clearly, but breathing seemed to require a greater effort, and my movements were slowing and becoming more sluggish.

As the minutes dragged into hours, I found myself becoming awed by the view this higher altitude afforded me. Just to the north, lightning peeled off the clouds, wavered vertically and horizontally from the depths of the cumulonimbus, and ignited the sky in a burst of light. To the south, the vista was no less dazzling in the vast darkness of a lonely section of the southwest Pacific sky where the stars took on a brightness and clarity denied more terrestrially based stargazers. I marveled at this breathtaking display of the heavens and, in my new proximity, felt an adopted kindredness. What a difference a mere seven thousand feet made in relation to the billions of miles that separated us from our celestial neighbors.

I wanted to laugh at the mundane problems of the dark planet rotating below me. In my delirious mood, I began singing softly. Before long, however, I was belting out a tune at the top of my lungs. I was having a grand time singing and laughing over the slightest thing until we cleared the front and began a descent back down to ten thousand feet. I greedily breathed in the heavier air. With every labored breath, I felt the sobering effects oxygen has on one after five hours at that hypoxic altitude. I was physically drained from working my lungs overtime and was glad the weather was clear at Singapore so I wouldn't have to fly the instrument approach.

Singapore's Paya Lebar International Airport was a busy hub for the Far East with airliners of different nationalities approaching the city from every direction. The controllers gave their instructions in rapid-fire broken English and didn't have time to repeat or speak slower. Using a combination of doing whatever Bennett did and also what I suspected the controllers wanted me to do, I managed to land without creating a major incident.

When the customs officer came out to our aircraft, he seemed mystified at our presence and asked what we were doing here.

"We're delivering these aircraft to Anrite Aviation," Bennett told him.

"But there is no Anrite Aviation here," the customs agent said, clearly annoyed at our presence. "Are you sure you are at the right airport?"

"The right airport? Of course," Bennett said, and snickered.

"Anrite Aviation is the Piper dealer," I said, though now with less certainty. There were dozens of airliners parked on the ramp or taxiing to the terminal, but no small aircraft.

The agent had never heard of a Piper or Anrite anything. He told us to follow him to his office where he checked a list of the airport tenants, but could find nothing resembling either of those names.

I thought back to Moody pointing to the map in his office and telling us, "Take these two aircraft over here, to Singapore." Well, here we were in Singapore, where we were supposed to be, but nobody seemed to know what to do with us. Eventually, the officer checked with another customs official, who told us that Anrite Aviation was at Seletar Airbase, Singapore's general aviation airport.

"They must think we're a couple of bozos," I said, as we walked back to our aircraft to fly the short distance to Seletar. "We've just come halfway around the world and don't even know where we're supposed to take the airplanes."

Bennett, optimisic for a change, said, "At least we got here in time to make the Pan Am flight tomorrow morning."

"Tomorrow morning?" exclaimed Pat White, in an upper-crust British accent. "I don't think you'll have much luck doing that. Not if you want to fly on Pan American."

Tall, distinguished, white-haired, and in his early sixties, White looked like the plantation owner, the consul general, and the old duffer in the gentleman's club who always had a witty comment about the queen or the latest cricket match. He was the Piper distributor for South East Asia. In keeping with the image of British colonialists, his Indian secretary brought us a tray of tea and scones.

"What's happening with Pan Am?" I asked.

"They've cut back a number of flights because of this bloody Arab oil embargo. The next one to Honolulu isn't for three days."

After checking into a hotel, Bennett and I hired a taxi to take us downtown. We didn't know where exactly we wanted to go, just someplace where something was going on, we told the driver.

"Oh, girls, girls. You want girls," the cabdriver said, smiling and shaking his head in understanding or maybe longing. "I take you where you find action."

Where I figured the action was would be a smoky nightclub with a sexy cabaret singer, uninhibited, mini-skirted young ladies, a few high-class call girls discretely plying their trade, and at least one fat guy in a white suit making shady deals at a corner table. The cabdriver, interpreting a somewhat different meaning, took us to an old house in a run-down part of the city and told us to wait in the parlor until another older Oriental man joined us. He and our cabby exchanged several words in Chinese, while Bennett and I waited, not sure what was going on.

"You think we're going to get shanghaied?" I whispered to Bennett.

"I don't know, but I can just see Walt now, asking Phil 'Whatever happened to those two guys I sent to Singapore?' "

A few minutes later an old woman came into the room, followed by several young girls who, giggling and whispering among themselves, lined up against the wall. Then silence, as the cabby, the old man, the old woman, and all the girls looked at the two of us.

Since Bennett and I weren't doing anything except looking at each other stupidly, the old man, translating for the old woman, said, "Young girls. Virgin. You like, yes? You pick now."

Picking any of these girls was something that neither Bennett nor I had in mind. We extricated ourselves as delicately as we could and left with the cabdriver, who, disappointed at losing his commission from the brothel, kept muttering under his breath as he drove us back to the hotel.

Called the Lion City, Singapore is a mix of Malaysian, Chinese, and Indian culture thrown together to form a rich montage of colors, sounds, and smells. Only nine years had passed since its independence, and the city still had the look and feel of a proper British colony with its low, flat skyline and quaint, tree-lined streets. All that, however, was changing. Western progress was slowly, but inevitably, encroaching on eastern tradition. Orchard Road, the tourist and shopping paradise in the heart of the city, was one massive construction site with girders, cement mixers, and the ever-present din of jackhammers. With nowhere to build but up, it wouldn't be long before Singapore looked like any other big international city with the glass, chrome, and concrete of characterless skyscrapers, megalopolis five-star hotels, and, the ultimate symbol of rampant capitalism itself, McDonalds.

Our aimless wandering eventually brought us to Singapore's single most important tourist attraction, the stately Raffles Hotel. Built in the late nineteenth century and named after Sir Thomas Stamford Raffles, the East India Company executive who founded Singapore, the hotel was one of the last vestiges of a bygone era. In its prime it had hosted such literary giants as Noel Coward and Rudyard Kipling, along with British government officials, foreign agents, traders, adventurers, and spies, rumored to have exchanged secrets in the Moorish-looking lobby bar, where more than sixty years ago some anonymous bartender mixed a concoction of spirits and created the Singapore Sling.

We sat at a table to have one drink. Eventually we forgot all about exploring the city as our thoughts turned once again to adventure. The second or third drink may have helped, but it was also because of the guy at the table next to ours.

"You guys from K L?" he asked, his eyes glancing toward the lobby.

"What's K L?" I asked.

"Kuala Lumpor—in Malaysia," he said, and took a long drag from his Dunhill. "You look like you're new to Singapore."

"Just got here yesterday," said Bennett. "How about you?"

"Came in from Bangkok a week ago. Supposed to meet some people from K L, then I'm off again. Singapore's not a bad place, but they're a bunch of neat freaks. Throw a cigarette butt on the sidewalk, and they'll lock you up. What a way to run a country. Pre-war Saigon, now *that* was the place. Singapore's kind of like that—money, women, drugs, you name it. Plenty of expatriates too: Americans, Brits, even the Australians. You don't know who's who anymore."

"So who are you?" asked Bennett. "What're *you* doing here?"

"Ah, I just do what needs to be done," he said, then changed the subject and lowered his voice, as if he was telling us a secret that no one else should hear. That clinched it for me. The guy sounded like a freelance mercenary, CIA operative, arms smuggler, or maybe a disillusioned husband who decreed himself a divorce-by-logistics. The way he kept looking over his shoulder made me think he might be hiding out from some drug lord whose thugs were at this moment lurking behind the potted palms. He was a Westerner, but had obviously spent several years traipsing around Southeast Asia and at some point probably felt it advantageous to jettison any traces of nationality. He could have been English, German, Australian, or from Des Moines, Iowa. He regaled us with stories of his nefarious escapades in the Far East until it probably dawned on him that he didn't know if we were the good guys or the bad guys.

"Where you guys from, anyway?" he asked suspiciously.

Bennett was up for the game. Either he was being just as vague or he had convinced himself that he actually belonged in the small outback town we left two days ago, because he answered, without skipping a beat, "Mount Isa, Australia."

"Nice place," said the stranger, sneaking another furtive glance toward the door before continuing. "Passed through there in sixty-nine, when it was just a podunk little town with more kangaroos than people."

This reminded him of yet another story, which he proceeded to share with us. I didn't know if he was making it all up or just exaggerating, but one thing seemed real, he had spent time in the Far East. So we listened with interest as he talked of adventure, intrigue, and, of course, colorful characters. It was a distinction I had not realized until now. I had started out on this trip looking for adventure and found people: Cary Wade, the expatriate American on Pago Pago who kept his sanity by acting a little crazy; the tired waitress in the outback town of Mount Isa demanding courtesy; Pat White, the caricature of British colonialism; and the cabdriver who moonlighted as pimp.

If Singapore really was a place where Old China Hands hung out, looking to either make their fortune or hide from all the cutthroats and bounty hunters that wanted to kill them, then this teller of tales was home. In time, Bennett and I decided to continue our exploration of the city. We left the mysterious stranger exactly as we found him, keeping vigil from his perch in the bar and looking over his shoulder every few minutes.

5

Teaching the Ropes

"It's about time you got back," Dave McCollum said. "We were beginning to think you and Bennett got shanghaied."

"There were a few times I thought that was a distinct possibility," I said, thinking back to the unexpected visit to the whorehouse. I handed McCollum the export declarations stamped by U.S. Customs as proof for the Commerce Department that the aircraft left the country. He filed it away and returned to his desk, strewn with a wide assortment of official-looking papers. The telephone in front, telex machine to the right, filing cabinet behind him, and typewriter on his left surrounded him like a strategically placed defense perimeter.

It was McCollum's job to make sure all paperwork was completed on every aircraft Globe Aero ferried. Each airplane required close to a small mountain of documents that had to be correctly stamped, signed, filed, copied, notarized, and mailed before it was exported. McCollum was good at this necessary but underappreciated task. A paper foul-up was so rare that his contribution to a successful delivery was often taken for granted. Sadly, however, should even one aircraft be delayed because of some administrative oversight, his reputation would be tarnished for days.

"We got anything going to Europe?" I asked.

"What, are you bored with the Pacific already?"

"I'd just like to go someplace that has a little less ocean between it and Lock Haven. Besides, I want to see what the North Atlantic is like while it's still winter."

"That's a switch. Usually everybody's griping about the Atlantic. This is the first time anyone's ever said they couldn't wait to go. I don't know about Europe, though; I think Walt's got something else planned."

Just then Moody stepped out from his office. "Come on in," he said, and then shook my hand in what appeared to be sincere congratulations.

"That's two trips, now," he said, smiling, beaming actually. "So how do you like ferrying? Phil and Bennett said you did a good job."

"All I did was follow them. Anyone can do that."

"Well, they both said you learned fast, and you're ready to go out on your own."

Moody's praise was so far removed from his usually gruff demeanor that I was caught off guard. Sure I knew how to plot a course and had learned how to navigate over thousands of miles of ocean to a tiny island without getting lost. And for the most part, I had adapted to the rigorous demands of ferrying small aircraft and could get by in foreign countries. It seemed I had passed the test and, based on the obviously glowing accounts of my performance, I could be trusted to fly a trip by myself.

All it really meant, however, was that I didn't talk back to either Waldman or Bennett, I knew how to find longitude and latitude, and I was able to keep the airplane right-side-up while pointed somewhere in the right direction. Also in my favor was the fact that I didn't act like an idiot overseas and bring disgrace to Globe Aero or to my country.

Important as these factors were, I suspected the real reason Moody was happy was because I didn't quit and leave either aircraft in San Francisco. Pilots who didn't mind the risks or the long hours were hard to find, and right now what Moody needed most was pilots. With several aircraft parked outside the hangar and more waiting to be picked up at the factory, he was anxious to send me on another trip.

"Are you ready to go out again?" he asked.

"I guess so," I said, trying not to appear too eager. "Where am I going?"

“I’ve got to get these two Beechcraft to Australia. When do you think you can leave?”

“In a few days I suppose. Who’s taking the second airplane?”

“You’re getting to know the Pacific pretty good, so I’m going to send Holmes with you.”

Don Holmes was one of the new pilots hired a month before me. Because he had little piloting experience before coming to work for Globe Aero, Moody kept him flying around the country, picking up airplanes, and ferrying them to Lock Haven until he had enough flight time to be trusted on an overseas trip.

“I didn’t know Holmes was checked out over the ocean,” I said.

“He’s not. That’s why I want you to lead him.”

“Me? I hardly know enough about the ocean to keep myself out of trouble. I’m probably not the ideal person to be training anyone. Can’t you get someone else with more experience to lead him?”

It was a useless argument. Once Moody assigned someone to a trip, even if it was in thought only, it was next to impossible to get him to change his mind.

Aside from being disappointed over not getting to tackle the North Atlantic for yet another trip, I was still flattered to be trusted enough to lead a new pilot after having made only two trips myself. I felt a new elevation of stature and probably a little cockiness, even though I suspected Moody wanted me to lead Holmes simply because there was no one else available. Mindful of my new responsibility, however, I attempted to train Holmes as thoroughly as Waldman had trained me.

When Holmes came into the hangar later that day, we went over the charts and personal gear he would need for the trip, after which I explained about plotting courses and what little I knew of ocean navigation. He seemed eager to learn and accepted my hesitant teachings with no arguments. We worked out a rough itinerary and, thinking the rest of his training would best be done en route, I told him to meet me in San Francisco in three days and went home to pack.

Home, if I could call it that, was the Fallon, a faded old hotel on the banks of the Susquehanna River, which had seen better days. Like everything else in this small town that time seemed to have passed by, this hotel was neglected. The rooms smelled moldy, like a damp cellar, and it was known as the "Falling Down Fallon." But it was convenient and didn't require a lot of housekeeping on my part. As with the other Globe Aero pilots, convenience was an important factor in choosing where to live. Boarding houses, staying with friends, and basically living out of the trunk of one's car were the primary options, which is what I had been doing until Flournoy suggested the Fallon.

"Ask for a room in the old section," he had told me in Honolulu, when we were discussing Lock Haven's scarcity of available housing.

"What's so special about the old section?" I asked.

"It's cheap, that's what. The rooms were under water during the flood and rent for only four dollars a night. They may smell a little musty, but what do you care? You're never gonna be there."

Flournoy was apparently right. I had been back only one day and was already packing to leave again. Right after breakfast, I left my key with the front desk, told the clerk I would be back in two weeks, and took off for the West Coast.

Holmes was waiting for me three days later when I landed in San Francisco. He had arrived a day earlier and, now, excited and rearing to go, greeted me at the airplane as I taxied onto the ramp. "I checked on the winds with Redwood City this morning," he said, "and they said it looks good for tonight—a minus three and a minus eight at worst."

A minus three and eight referred to the winds at five and ten thousand feet respectively—which meant headwinds, but light enough that they shouldn't pose a problem. "That doesn't sound too bad," I said. "We can leave around three in the morning and still be in Honolulu by midafternoon. If we go to the hotel now, we might even be able to get eight hours sleep."

"Sleep? We can't go to sleep. We have to go to North Beach."

"Excuse me? How did North Beach get into the picture?"

"Well, we can't come all the way out here without seeing the place."

"That'd be a good idea, except that it's going to take about thirteen hours to fly to Honolulu. We should try to get some sleep—especially since this is your first crossing."

"But I got plenty of sleep last night. Besides, I've never been to San Francisco. Why don't we just go and look around. North Beach is supposed to be the big place here. Have you ever been there?"

"Well . . . no," I said, thinking that mixing in a little sightseeing wouldn't be a bad idea. "But we have to take advantage of these winds. If we don't cross tonight, who knows how long we'll be here."

"The briefer says the winds should stay good for a couple of days. And I've always wanted to see North Beach."

"If we go out tonight, we can't do any drinking, and we can't stay out too late."

"We'll just look around, check out a few nightclubs, and maybe have a drink or two, but I'm not going to get drunk. We can still get back to the hotel in time to get some sleep."

So we went to North Beach. We didn't get drunk, but it was close to midnight before we got back to the hotel. Four hours later, sober but tired, we dragged ourselves out of bed to fly to Honolulu.

The first several hours were the toughest. I fought a constant battle to keep awake, as I'm sure Holmes did. By the time we passed over Ocean Station November, the drowsiness had run its course, and I was wide-awake. I looked out my side window at Holmes's airplane, still in the same position he had been in since we left San Francisco, so I guess he, too, was able to ward off the sleepiness. But something wasn't right. He had kept up an almost constant monologue since takeoff, yet for the last thirty minutes the radio had been quiet. I tried calling him several times to see if everything was okay, but got no response.

It was possible his radio had quit, though not likely since it was brand new. Still, as Murphy's Law dictated, anything that could go wrong, would go wrong, and at the worst possible time. Being over the ocean on one's first crossing, or even on one's third crossing, was just such a time.

Fortunately, the aircrafts' standard VHF radios were not our sole means of communication. Like the navigation radios, they were limited to a few hundred miles at best, nowhere near enough to meet the requirements of oceanic communications. Air traffic control required us to report our position once every hour, and for that we used high frequency (HF) radios.

The ones Globe Aero had installed were basically ham radios hooked up independent of the aircraft's avionics to an antenna system that, if one were extremely generous, might be described as archaic. The antenna was nothing more than approximately one hundred feet of copper wire wrapped around a hand-cranked reel and fed into a standpipe drilled through the cockpit floor. At the end of the wire, sticking out from the belly of the airplane, was a funnel, weighted to keep it taut in the slipstream. The pilot was then able to reel out the antenna and change its length, depending on the frequency he was using.

The amazing thing about this entire setup was that it actually worked—most of the time. The radios probably weren't state-of-the-art technology when Moody used them to ferry his first airplanes more than a decade ago. What made it worse was that every last one of them had been under water during the flood. Rather than junk them and invest in new equipment, Waldman took the radios to Moody's house and dried them in his oven. Although the radios worked if atmospheric conditions were just right, more often than not, what we ended up with was a heavy, useless piece of junk that we had to lug around airports being careful not to cause any more damage.

It took every four-letter word I could think of and a sore arm from reeling the antenna in and out several times, but I finally got the radio to work. It wheezed and crackled until I adjusted the antenna to the right length and was able to call Holmes. Still there was no answer. It was a long shot anyway. I was now left with two possibilities: either every one of Holmes's radios had quit or he had fallen asleep.

I ruled out the former because we had talked about this during the preflight briefing. Even if Holmes didn't remember, I'm sure he would've given me some kind of signal, like flying in front of me and wagging his wings. He did nothing like this, so I assumed it was the latter. With the little sleep we had gotten before takeoff, this was likely what had happened.

As problems go, this was not serious. With the long hours, the boredom, and the solitude some sleep aloft was inevitable, no matter how much rest we got before taking off. I did it myself on my last trip, and most of the other Globe Aero pilots also admitted to falling asleep at one time or another, though, usually five- or ten-minute catnaps at most. Obviously, if the pilot is asleep, there's always a risk of colliding with another airplane, but the odds of that happening weren't as great as one might think. Most of the aircraft over the ocean were airline or military jets flying above thirty thousand feet. At the lower altitudes we flew, traffic was so rare that the chances of crashing into another airplane were so small we didn't even consider them. The problem was when you slept longer than a few minutes, as Holmes was doing. I didn't know how long he had been sleeping, but I did know it was longer than a catnap.

I throttled back and let his airplane overtake mine. I wanted to be in a position where I could keep an eye on him. Because his airplane was on autopilot, it wouldn't just start spinning out of control and plunge into the ocean, but I figured we should stay together. Our flight plan called for a six degree heading change in fifteen minutes. I was confident he would wake up by then.

I kept calling him, shouting into the microphone, and looking for anything I could use to make a loud noise. I banged a metal flashlight on the metal seat frame, while keying the microphone at the same time, but still he didn't respond.

It was getting closer to the time we had to make that heading change, and I had run out of ideas. If only I had an alarm clock. That would be loud enough and annoying enough to wake him if his radio was on and if he hadn't turned the volume down. I didn't have one, but maybe one of the airliners within range of us did. I knew I would sound like an idiot if I tried to explain to some stranger why I needed an alarm clock, but it was all I could come up with.

"Anybody listening on one-twenty-three-forty-five?" I announced on VHF over the common ocean frequency. "This is November-Four-Four-Two-Four-Whiskey."

Twenty seconds passed before someone answered. "November . . . something," a deep, steady voice said over the speaker. "This is PSA Seven-Six-Eight. Can we be of any help?"

"Uh . . . PSA, from Two-Four-Whiskey. This may sound kind of bizarre," I said, embarrassed. I tried to sound professional as I explained the situation and asked if they had an alarm clock or anything with a loud bell.

"An alarm clock? Hmm. We don't get many requests for something like that, but I'll see what we can do. You must be one of those little guys down there." I could imagine the airliner's pilots chuckling between themselves. If they were amused, at least they kept it to themselves and didn't make me feel any sillier than I already did.

One minute later, the airline pilot called on the discrete frequency Holmes and I had been using before he fell asleep. No alarm clock. No bell. He simply announced to Holmes in this calm, steady, airline-captain voice that his buddy was trying to call him. Well, that was a big help, I thought. Didn't he think I had been doing the exact same thing?

"November flight, from PSA Seven-Six-Eight. We can't seem to raise him. Anything else you'd like us to try?"

"No, I'm sure he'll wake up eventually. But thanks anyway."

"Okay. Good luck down there."

Time had run out. I needed to make a decision that I was too inexperienced to make. I could either stay with Holmes while we both got lost or change my heading as the flight plan directed—then at least one of us would be on course. Waldman had stressed the importance of holding to the flight-planned heading when he trained me, and I guess it made an impression, because that was the course of action I took.

It was daylight. As I turned to the new course, I watched Holmes's airplane becoming smaller and smaller until it was just a speck in the distance. A minute later it disappeared completely.

Ferry pilots are an independent breed—ill at ease in a flight crew, where the lines of command are strictly followed, most of us would be deemed too insubordinate if fortune placed us in that environment. The aircraft we ferried were single-pilot, and, as such, each pilot was the captain and operated the airplane as he saw fit. Although good sense prescribed taking advice from our more experienced colleagues, it was in no way mandatory. As lead pilot, though, I was not only responsible for Holmes's training but also his safety. I could have been more insistent about getting enough sleep in San Francisco, but I wasn't. Now we—and especially he—would suffer for it.

Several more minutes of silence passed before a sheepish Holmes finally asked over the radio's speaker, "Is anybody there?"

"Well, good morning," I said, glad that he finally woke up but also angry that he had slept for so long. I tried not to sound too sarcastic, though, because I was sure he felt bad enough.

"I don't have you anywhere in sight. What's going on?" he asked. He sounded as if he just woke up alone on a different planet with nothing but water in every direction.

What's going on, I thought, a little angrily, is that you fell asleep for almost an hour. Instead, I said, "We had a heading change about twenty minutes ago, and I figured one of us needed to stay on course."

"What do you want me to do? Should I change my heading too?"

"Let me think about this for a minute."

Think about what? I wondered. We were in the middle of the Pacific Ocean, nine hundred miles from Honolulu and the pilot I was responsible for was boring a hole in the sky that might or might not take him anywhere near the Hawaiian Islands. It was conceivable that missing one small heading change wouldn't affect his course that drastically and that he would be able to find Honolulu by himself. But at what cost? I know if I had gotten separated from Waldman on my first crossing, I would've worried for the rest of the flight.

There had to be something I could do. Fuel wasn't a problem, although it would be if we started flying around in circles trying to find each other. I racked my brain for the solution and then remembered something Waldman had told me during my first crossing.

"Remember, these are only forecast winds," Waldman said, "which may or may not be anything like what you're actually flying in. You might very well have done all your plotting as good as anyone, and still be a hundred miles or more off course. And if your nav radios quit, you could shoot right past the island and never see it. That's when the squelch on the comm radio comes in handy. You probably didn't know you could use it to navigate by."

I didn't. I had never heard of anything so far-fetched. So I listened as Waldman explained the procedure in detail: the angles, the degrees, the miles, and the mathematical formula used to compute all these variables. He further explained that one of the radios, specifically the one that was transmitting, should be stationary—unlike Holmes and I, who were both moving. I also remembered his answer when I asked if he had ever navigated that way himself.

"Well . . . I haven't actually tried it. But if you're lost you'll, want to try anything you can think of to keep from going swimming, because if you're at your ETA and don't see anything, by that time, what have you got to lose?"

That time for Holmes and me was now. I made some quick calculations: the width of one degree in sixty miles was one mile. Holmes and I got separated twenty minutes ago, which, at our speed, meant we had traveled fifty-three miles. Close enough to call sixty. One mile times six degrees of heading change came to six miles. Not very far, and if it was still dark out I would probably be able to see his position lights even at that distance. In daylight, however, he could be half a mile away, and I still might not be able to see him.

We were both moving at the same speed, so I had to assume he hadn't passed or fallen behind, but was directly abeam of me.

"Turn right thirty-six degrees to a heading of two-six-one and hold that for ten minutes," I said. "I should be on your right, so keep an eye out."

Once each minute I had Holmes transmit. I turned the squelch down on my radio to where I could just barely hear his voice. Fully open, it received transmissions in a wide area, and by turning the squelch down I was narrowing that area. Each time Holmes transmitted, I kept turning down the squelch. I had no way of telling how far away he was, but at least I knew he was getting closer.

After ten minutes of decreasing turns, I had my squelch turned down almost as far as it would go and told Holmes to hold that heading for another minute before transmitting again. I said, "If I don't answer at that time, make a left turn to a heading of two-two-six degrees and transmit continuously."

One minute later, not having changed the squelch since we last talked, the radio remained silent. Now, everything hinged on whether or not I would be able to hear him again at precisely the right time. Holmes should be about two and a half miles to my right and starting a turn to intercept me. At a five-degree interception angle, I estimated he should cover that distance in thirty seconds. I waited with my fingers crossed.

". . . four, three, two, one. Do you read me?"

"Loud and clear," I said, becoming more amazed every time I had Holmes do something that worked exactly as it was supposed to.

"Now hold that heading for another twenty seconds, then turn right to a heading of two-three-six degrees."

I had him zigzag like this, decreasing the headings by one degree on each turn, until he was flying the same heading as me. With my squelch turned down as far as it would go I was able to hear his transmissions. He had to be almost on top of me, I thought, leaning forward and stretching my neck so I could look out the top of the windshield. Crashing into the only other airplane over the largest expanse of water in the world would be difficult to explain to Moody—if I was alive to explain it.

I spotted Holmes's airplane a minute later about five hundred feet off my left side and slightly behind me, which is where he stayed until we landed at Honolulu. From then on, whenever I suggested we get some sleep, Holmes didn't argue. Whenever I looked out my left window, Holmes was there, keeping a safe distance, yet close enough to easily be seen. And he was always awake.

All in all, this was a good incident to happen so early in Holmes's—or any pilot's—career. For nothing that I or anyone else could say would've adequately stressed just how big the Pacific Ocean really was, how lonely it could get when there was no other airplane in sight, and how important it was to get enough sleep.

But if Holmes learned anything from this incident, it didn't make a lasting impression. After a few months, we began to notice a peculiar pattern to his flying: whenever he left for the Pacific, he would fly directly to San Francisco, refuel, and leave right away for Honolulu. This in itself wasn't that strange because weather more than anything determined our pace, and flying thirty or so hours without rest was sometimes necessary because of winds. The odd part was that after rushing to Honolulu, he stayed several days before continuing on to Australia or wherever he was supposed to deliver the aircraft.

He was also going out to the Pacific more often, at least more than any of the other pilots. The few Atlantic trips he did take were quick ones—usually a day or two quicker than anyone else might take. And no matter where he went, he always seemed to get back just in time to take the next Pacific-bound airplane. The only time we saw him in Lock Haven, he was either coming off or going on a trip. We began calling him "Here Today Here Tomorrow."

Though it should have been obvious from the beginning, we didn't learn the reason for this strange behavior until several months later. On one of his early trips in the Pacific, he had met a young lady in Honolulu and romance was now in bloom. As so often happens when we're in love, the sound judgment of an uncluttered mind is often overruled by the desires of a much lower organ. No longer satisfied with the few days he and the girl spent together on the outbound leg of his journey, Holmes began flying from Honolulu to Australia at the same pace. He would make a quick refueling stop in Pago and again in Nadi, deliver the airplane, and be on a flight to Honolulu that evening. Then he stayed for several more days before getting back on an airliner to fly to Lock Haven.

"I want to see you take these trips a little slower," Moody told him on several occasions. When that had no effect, several of the pilots tried to get him to slow down. Even the girl's father tried, though his reasons were different—he just didn't like Holmes. The place he chose to make sure the young pilot understood this was the International Arrivals Hall at Honolulu Airport. As soon as Holmes got off the flight from Sydney, the father confronted him, telling him, with words and gestures bolstered by alcohol, to stay away from his daughter. For his own safety, airport security had to escort Holmes to his connecting flight.

That slowed him down for a while, but then he either won over the father or they declared a truce, because Holmes went right back to his old schedule and was soon flying almost exclusively in the Pacific. When he left on a trip his destination was Honolulu, and after he delivered the airplane his destination was Honolulu. Lock Haven was just a place he stopped to collect his mail, pack a change of clothes, and get his next trip assignment.

Pacific trips were hard when we flew them at a normal pace. There were too many miles and too many time zones. One needed a few days just to let their internal clock catch up with local time. Though the rest of us were pushed at times to fly a fast trip, we made plenty of leisurely ones. Holmes made none, and the toll it was taking was obvious during the few times he was home long enough to join several of us for dinner. Before we could get through the entree, Holmes would be fast asleep with his head resting in a plate of half-eaten food.

At the time we thought the whole thing was amusing—all this rushing around and pushing himself because of some romance that would probably be over in a few months. Although every one of us had at one time or other been guilty of the same thing, our jokes about his odd sleeping habits went on constantly, and not even the most reckless among us could foresee the fate that awaited him in years to come.

6

The North Atlantic

Ron "Doomsday" Mills handed Bob Campbell and me each a blue folder containing the weather charts and forecasts for the North Atlantic and turned around a hand-drawn colored map of the ocean lying on the counter so that it faced us.

"You want to go to Shannon, eh?" said the weather briefer, in a voice hardly more than a whisper. "Well . . . I don't think that's such a good idea right now."

"You don't!" said Campbell. "Why? What's going on out there?"

Mills shuffled some papers aside and leaned forward to point out the weather we would have to fly through. "There's this front at mid-ocean here, you see. And the freezing level's on the surface at least halfway across," he said, almost apologetically, as if embarrassed to be the bearer of such bad news. "It starts to rise a little on the other side, but not more than a few thousand feet."

"What about the front? Is there much to it?" Campbell asked. We both leaned closer to hear the briefer.

"Well, these satellite pictures indicate quite a bit of convective activity, and we've gotten reports of hail from several ships. So there'll probably be some lightning, and I'd expect a fair amount of icing once you hit the clouds." Mills paused. "You're not really serious about crossing tonight are you?"

Campbell studied the charts and the high-altitude satellite picture before announcing his decision. "I guess it looks pretty good. Let's go ahead and plot these winds and get going."

I looked at the charts again to see if I had missed anything. Although I didn't know as much about North Atlantic weather as Campbell did, I knew the tightly drawn lines and blackened circles with a lot of scribbling next to them meant lousy weather. But even more ominous was the briefer, whose tone seemed a clear warning.

"I kind of got the impression from the briefer that it would be suicide to cross with this weather," I said, after Campbell and I went downstairs to the restaurant to plot the winds.

"That's why they call him 'Doomsday,' " Campbell said, in a cavernous voice that came from somewhere deep in his gut. "A while ago he briefed a pilot who went down over the ocean after picking up a load of ice and never came back. Ever since then he's become so conservative with his briefings that everybody started calling him 'Doomsday.' So if you have any hesitation at all about going, Mills will talk you out of it."

"What about this freezing level on the surface, though? Looks to me like that's something we *should* be worrying about."

"But we'll be on top most of the time. And when we do hit the front," he said, while pointing to the chart's contoured lines, "see how the freezing level rises quickly because of the warm air from the Gulf Stream. Even if we do pick up ice, by that time we'll have a few thousand feet of warm air under us."

"That's not much."

"Don't forget it's still winter up here. There's no way you can get across this time of year without picking up some ice. There's almost always going to be a front out there somewhere. But as long as you have a freezing level, even if it's just a thousand feet, you'll have some room to get rid of the ice."

"You have to know what the weather's doing," he cautioned, as he brushed aside an unruly shock of thick, black hair that kept falling over his forehead. "Keep track of where the lows and fronts are, and try to time the takeoff so it'll be daylight when you hit the front. It's a lot easier to pick your way around the clouds if you can see them. And always have a plan for when

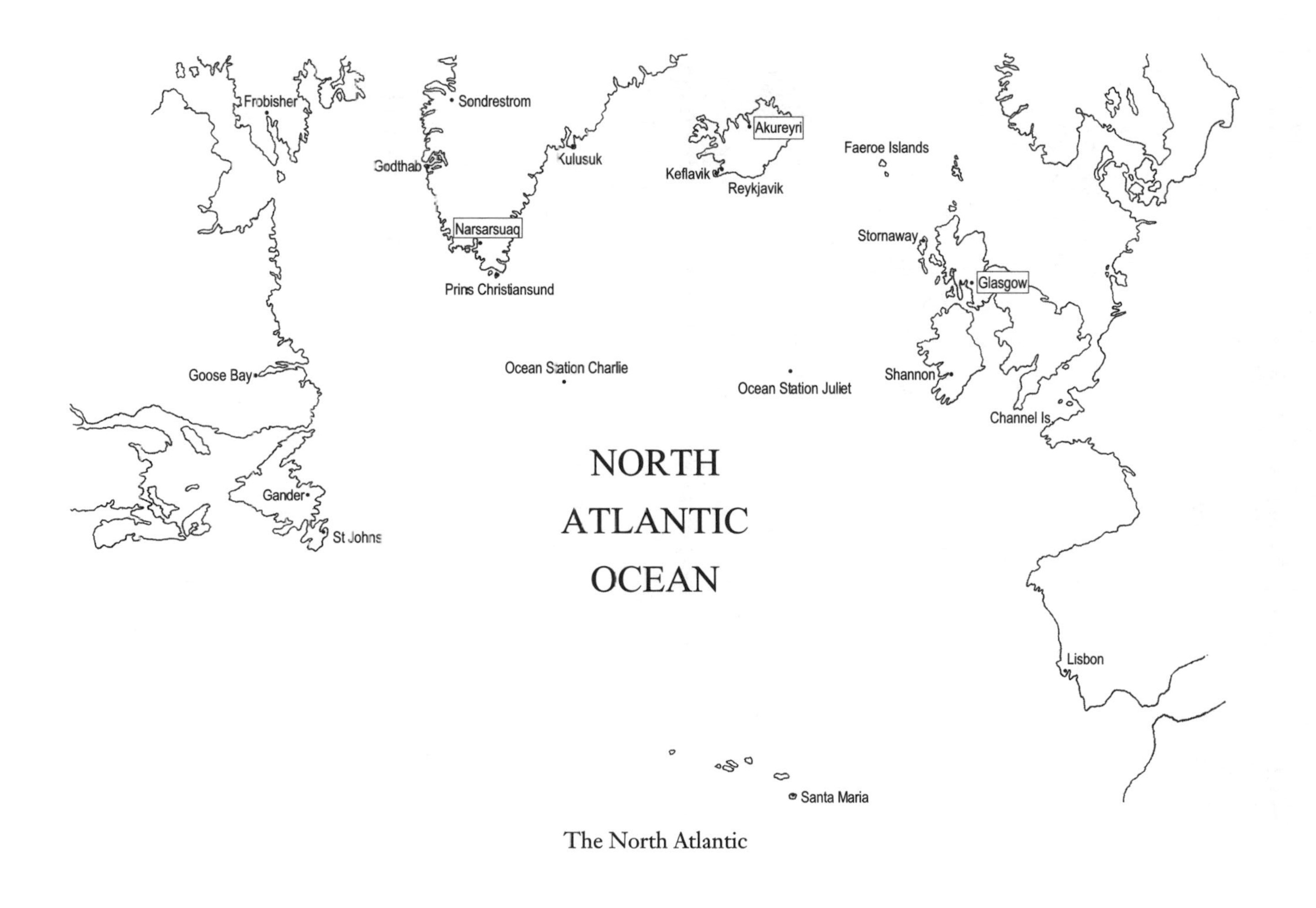

The North Atlantic

everything turns to crap. Because if you come out here enough, sooner or later it will."

I listened to Campbell's advice with the rapt attention of a novice. I had heard stories about the treacherous weather in the northern latitudes of Canada and the North Atlantic, but they were just that . . . stories. Until now, my experience with winter flying had been limited to the relatively tame weather of the northeast United States, where, if I encountered bad weather I could always turn around, or choose another route, or simply duck into a nearby airport and wait it out. Over the ocean those options weren't available, and if I got into trouble, there wasn't much I could do about it.

Campbell's explanation of the weather was pretty much the same as the briefer's. At least all the facts were the same. Coming from Campbell, however, it didn't sound anywhere near as bad. Besides having several years of experience over the ocean, he was the only pilot at Globe Aero—probably the only ferry pilot anywhere—who knew the ocean from both in the air and on the sea. Before getting into ferrying, he had spent fourteen years in the U.S. Coast Guard as a radio operator on ocean station ships in the North Atlantic.

"My enlistment was almost up, and I was tired of spending a month at a time at sea," he told me. "I'd gotten friendly with Walt when he used to call in his position reports to the ship, so one time I just asked him for a job."

"Wait a minute. I must have missed something, because I don't see how you go from operating radios to flying airplanes."

"I'd been taking flying lessons while I was on shore duty back in Portsmouth and had all my licenses. So when Walt told me to come up to Lock Haven and he'd send me out on a trip, I took some leave and followed him to the Philippines. As soon as I got back, I took my discharge and went to work full time for Globe."

In the four years since, Campbell had made numerous crossings over both the Atlantic and the Pacific and was now Globe Aero's chief pilot. I felt fortunate to be able to rely on his experience for what would undoubtedly be a real test of any pilot's skill.

If nothing else, at least we had good airplanes: two twin-engine Aztecs we were taking to the Piper distributor in Genoa, Italy. I was beginning to feel more confident of our success, until we walked outside to the aircraft. A biting wind howled across the ramp, unsettling the mounds of snow piled high to the side of the taxiway. The cold, damp, night air had chased everyone inside, making the airport look lonelier and more barren than it probably was. In so remote a place at the easternmost tip of the North American continent, it felt like you were at the very edge of the Earth. There was nothing east of here except seventeen hundred miles of dark and stormy ocean that meted out punishment swiftly and without favor.

While the Pacific Ocean was a benign giant where the boredom was occasionally interrupted by an isolated cloud formation, the Atlantic was just the opposite: a mixed soup of cloud, rain, and snow, stirred by the arctic winds of the jet stream and the warm waters of the Gulf Stream. There were deep low-pressure areas with mountainous cloud formations spinning counterclockwise as they hulked across the ocean. And there were cold fronts spawned off these lows, carrying their cargoes of hail, turbulence, and ice, and waiting patiently to vent their anger on anyone foolish enough to enter their domain.

I kept Campbell's airplane in sight during the takeoff and climbout, leveled off two thousand feet below him, and adjusted power to close the distance between us as soon as we turned on course. I figured Campbell knew what he was doing and as long as I stayed with him, our fates—to whatever degree they were affected by the weather—would be identical. It was clear over Newfoundland, and, as we approached the coastline, I saw the scattered lights of the small village of Wesleyville, which announced a too-hasty farewell to the North American continent, and then disappeared as we droned on toward Shannon, Ireland.

At first the still void of ocean and black sky merged into one. That was soon interrupted by the eerie glow of the Aurora Borealis on the northern horizon. A few miles to the southeast I saw the flickering lights of an eastward-moving ship slowly cutting along the surface of the water. I envied the seamen for their ability to get up and move about the deck, even though my crossing would be over in less than thirteen hours and theirs would take several days. I was also in a warm cockpit, and they were getting drenched in rain-soaked seas and tossed around in thirty-foot swells while sailing through a storm.

The relative comfort I enjoyed was not without its price, however. Should some disaster befall either craft, I suspected the seamen's chances of survival were somewhat greater than mine. They had large, forty-man lifeboats, flares, portable radios, and probably enough emergency provisions to last a week or longer. If I were to ditch, I would have to survive in a two-man life raft for God knows how long on a quart of water and a few candy bars. That was if my life raft even inflated, something I hadn't thought about until Campbell asked if I had checked the date it was last inspected.

"No," I said. "When was the last time you checked yours?"

"I don't know. I don't carry one in the winter."

"You must have a lot of faith," I said, surprised that anyone would even consider flying over the ocean without at least a life raft. "Either that or you're an awfully good swimmer."

"You have any idea how cold the ocean is? Because unless search and rescue is right there to pick you up as soon as you hit the water, you're not going to last more than a few minutes before hypothermia sets in."

"So what'll you do if you have to ditch?"

"Well, first of all, I try not to think about it. But if I lose an engine or pick up too much ice and the airplane starts heading for the water, I'm just going to point the nose straight down and get it over with quick."

That sounded like a damn-the-torpedoes-and-full-speed-ahead! thing to say, but I could understand his reasoning. Even with a life raft, the chances of surviving an ocean ditching aren't even worth calculating.

What was I really doing here? Was it money, or some sense of adventure that had caused me to look forward to this crossing with such eagerness? If so, I was not alone, for since the early part of the century there has been no shortage of pilots obsessed with the challenge of pitting their intelligence, skill, and strength against this formidable ocean. They came first with bal loons and rickety flying boats and then airplanes, all seeking the fame and glory that such a feat would bring. While history remembers the lucky ones like Charles Lindbergh, John Alcock and Arthur Brown, and even Douglas Corrigan, who succeeded, there were many others who crashed or left the shores of North America and were never seen again.

I hesitate to include myself in the same company as those pioneers, for, although my airplane is small and as flimsy against the might of nature as theirs, it is better equipped, with modern radios and flight instruments, and engines considerably more reliable. Of even greater importance is that Campbell and I aren't flying into the unknown, but over the most heavily traveled air route in the world.

I was startled out of my daydream by a stark, light-gray puff of cumulus that washed over my airplane for a second or two, obliterating everything in sight, and then disappeared as quickly as it came. It was still dark, but I didn't need daylight to tell me we were approaching the front. And it was more than an hour early.

"How's the weather up there?" I asked Campbell.

"I'm still in the clear at eleven, but it looks like the tops are getting higher. How is it at nine?"

"I just went through a cloud and it looks worse up ahead. Either the front's slowed down or we've got one hell of a tailwind. What do you think?"

"It could be either. Gander's usually good at forecasting this stuff, but even they miss it sometimes."

"Yeah, well, it looks like the freezing level's still on the surface. They didn't miss that. I'm going to climb up to your altitude and see if I can get out of this stuff."

There were cloud buildups all around, but they were isolated enough that I was able to stay in the clear as I climbed to eleven thousand. I kept Campbell in sight, yo-yoing back and forth between a quarter of a mile and a few hundred feet. The horizon, which hadn't been that easy to see to begin with, was now completely washed away. I was flying in haze or maybe between cloud layers; I wasn't certain because it was too dark to tell. So much for our plans of picking our way around the clouds in daylight.

I tucked my airplane in close to Campbell's, about ten feet behind his right wing, if I judged distance accurately in the muck. We were doing one hundred fifty knots, zipping in and out of clouds so fast it would mean instant disaster if one of us got careless. We were still several hundred miles from land in any direction, and I couldn't get anything on the radio. Not even Ocean Station Juliet, which was just four or five hundred miles in front of us. I could hear nothing on the radio except Campbell, and I could see nothing except the position lights on his airplane.

We penetrated deeper into the front, bouncing around in the turbulence and dense clouds that pounded the airplane with rain and blanketed everything in sight. I could barely see my wingtips, let alone Campbell's. I clutched the control wheel and throttles in a death grip, ready to dive out of the way if his airplane didn't reappear in the exact same spot I last saw it.

This was crazy. As much as I disliked the thought of losing sight of Campbell, I liked the idea of flying into him even less. "It's getting too bumpy up here. I'm going back down to nine," I told Campbell, as I reduced power and began to descend.

"That's probably a good idea. This front looks like it's a couple of hundred miles thick, and I imagine it's going to get a lot worse before we break out, so we might as well play it safe."

"You getting anything on ADF yet?"

"Not right now. I was receiving the Consolan station at Bushmills a while ago, but not good enough to navigate with. Too much static from the storm."

"Well, I don't have anything else to do right now except take a look at my wing every few minutes to see if I'm picking up any ice," I said, again looking out the window. "I think I'll see if I have any better luck."

Consolan operates in the low frequency radio band, emitting a continuous series of sixty dots and dashes that correspond to specific radials from the station. All one had to do was count the number of dots and dashes, and if the signal was strong, and assuming a rough approximation of position was already known, bearing from the station could be established. Although the system dates back to the German U-boat era, it was remarkably accurate and, to the trained ear of an experienced navigator, could be used for several hundred miles. For ferry pilots, with no way of determining our position other than a compass and a clock, it was considered to be technology of the highest order.

I tried both the Bushmills station in northern Ireland and the Ploneis station on the French coast with an equal lack of success. Too much of the signal was lost in the electronic clutter of the storm—I could neither count enough of the tones nor determine if I was hearing a dot or a dash. After a while, I gave up and concentrated again on my immediate surroundings. A glow of pale red dots—ice crystals or snow reflecting light from the red position light—was shrouding my left wingtip. In front was still the same blackness, but leaning forward and peering through squinted eyes I could make out the nebulous shapes of snow rushing toward me. I turned on the landing lights and was immediately bombarded with an infinite number of tiny dots of light.

As captivating as the display was, it was also hard on my eyes, so I turned off the landing lights, preferring instead, the dull blackness of night. A faint blue-green glow was emanating from the tips of the propellers and forming an eerie circle of static discharge—Saint Elmo's Fire, named after the patron saint of navigators. I knew it had some mystical significance, but whether it was a sign of good or bad luck I couldn't remember. Out of curiosity, and maybe a little superstition, I was about to ask Campbell if he knew what it meant when something of greater urgency caught my attention.

A thin layer of ice had begun to form over the leading edge of the wings. It was not enough to cause the airplane to lose airspeed, but I knew that as long as I stayed in the clouds, I would pick up more ice. Then it would only be a matter of time before the airplane began to slow down.

"I'm picking up some ice down here," I said. "How're you doing at eleven?"

"I've got a light trace, too, but it's not building very fast."

"We should have a one or two thousand-foot freezing level by now. You think we ought to start down?"

"If we do we'll lose the tailwind. Besides, the ice isn't much right now. Why don't we stay here for a while and see what happens?"

Twenty minutes later ice was still forming, and airspeed had dropped fifteen knots. The outside temperature was minus fifteen degrees Centigrade, which I figured should put the freezing level at about two thousand feet. That didn't leave a lot of room to melt the ice, but it would be enough if I didn't wait too long to descend. I called Campbell to ask if he thought now might be a good time to descend to a warmer altitude.

Campbell didn't answer. I called again, but got nothing. Could he have fallen asleep like Holmes did in the Pacific? I doubted it, but his radios could've quit, he could've had an electrical failure, or any number of things could've gone wrong. In any case, I didn't want to wait to tell him I was going to start down. Ice was forming in large chunks, and I had to do something.

I selected nose down trim on the autopilot, but nothing happened. I punched buttons and turned dials, and still no response. I tried to rotate the pitch-trim wheel. But it was either jammed or frozen, and I didn't want to force it for fear of breaking a control cable. Ice was still building on the wings, and I lost another five knots of speed in the short time I was trying to get the trim to work.

A fine time for the autopilot to go haywire, I thought. I pressed the disconnect switch and tried to push forward on the control wheel, but I couldn't get it to move. I applied more pressure, but it, too, must have been frozen because nothing happened. I could turn the wheel left and right and could work the rudder pedals, which meant I could still bank the airplane, but that wouldn't help. What I needed to do was descend to the warmer air below the freezing level, and I had to do it before the airplane lost so much speed that it quit flying altogether. Airspeed was down to one hundred knots and dropping rapidly when the idea came to me—a solution so basic I should have thought of it before. I just hoped the throttle cables weren't frozen, too.

I reduced power, and the airplane settled into a shallow descent. The temperature was getting warmer as I dropped down into the thick, dark clouds. But it was rising too slowly. At five thousand feet it was still minus seven degrees Centigrade, which would put the freezing level at about fifteen hundred feet. That was just the altitude where ice no longer accumulated. I would still have to descend another five hundred feet or so to where the temperature was on the plus side if I wanted to melt the ice. Because of the darkness and the clouds, I didn't want to descend too low if I didn't know what my actual altitude above the water was. And with no way of setting the aircraft's altimeter to the local barometric pressure, it could easily be wrong by five hundred or even a thousand feet.

I watched the altimeter unwind with increasing anticipation, as the temperature crept slowly upward. At seventeen hundred feet, the outside air temperature gauge read zero. At this point it was the most important instrument in the airplane, and I hoped it was accurate. Another six hundred feet lower and the ice began to break off, slamming against the fuselage as it melted from the propeller and was slingshot free. Large gaps appeared on the wing as sections of ice broke off. A few minutes later the last of the ice was gone, pitch trim and elevator controls were back to normal, and airspeed had begun to increase.

The Aztec no longer waddled through the air, but was once again aerodynamically sound and boring a hole into the darkness. I was no longer in clouds, but I might as well have been because all around me was ink black. I couldn't distinguish the water below from the clouds above. It was as if I had flown into a black hole.

Where was Campbell?

Even worse than the loneliness was that I didn't know how close I was to the ocean—a few hundred feet or a few feet. The only thing I was certain of was that because I was flying into an area of low pressure, my actual altitude above the water was lower than what the altimeter indicated. I longed for the psychological safety of the altitude I had left two miles above the ocean. I couldn't return to that altitude without picking up ice again, but I could climb back to the freezing level, which I knew put at least six hundred feet between myself and the ocean. Increasing power, I climbed back up to the freezing level and, deciding I wanted a little more insurance, kept right on climbing, though ready to start back down again at the first sign of ice.

"You still here?" Campbell's voice bellowed over the speaker.

Finally. I was overjoyed at the sound of his voice. "I'm still plodding away down here," I said. "I've been trying to call you. What happened?"

"I was on HF trying to get a position report out. Anything exciting happen?"

"Well . . . sort of. I picked up some ice and had to drop down to about a thousand feet to get rid of it, and . . . oh yeah, the flight controls froze up. Does that happen a lot?"

"Not really. Only on some of the Pipers. It's the grease they use on the pitch-trim cables. It freezes when it gets real cold."

"Like now, I guess."

"Exactly. Walt's been mentioning it to Piper for several years now, so at least they know about it."

"That makes me feel a lot better," I said.

"Welcome to the North Atlantic," he said, and gave a short laugh. "This is normal. What you should've done is go up instead of down. I climbed to thirteen thousand a while ago and it was too cold to pick up any ice."

Up or down. It was an age-old choice that pilots have had to make whenever they encountered a cloud layer. Go on top, which could get you into trouble if the clouds rose higher than the airplane was capable of flying, or go underneath, where you might run out of altitude if the clouds extended all the way to the surface. Either way was a toss-up; Campbell had chosen correctly this time, while I had not.

"Are you still in cloud?" I asked.

"No, I broke out about ten minutes ago. Why don't you climb back up?"

I passed through a ragged strata of clouds, at first dark and sinister and then lighter as I continued the ascent. There was no ice. Sunrise had come some time ago, but with the clouds blocking any light, it stayed dark until I broke out on top of the stratus layer. I had seen sunrises over the ocean before, but none more welcome than this one. The morning sky was so peaceful and serene, that it made the events of the night seem like a bad dream. Even the radio, free of interference from the storm, pointed straight ahead to Ocean Station Juliet. Campbell, of course, was nowhere in sight, but we joined up again over the ship and tracked in toward Ireland on the BBC until Shannon Control picked us up on radar and vectored us in for a landing on a glistening runway still damp from the late-morning mist.

Shannon Airport rests on the banks of the Shannon Estuary in County Limerick, forty miles inland from Ireland's western coast. The green rolling hills and swampy peat bogs surrounding the airport are the first glimpse of color—and, more important, solid earth—one sees after a long night over the drab gray of the ocean, and for this I felt immense gratitude.

If Ireland abounds in anything, it is castles and pubs. The former, which Campbell and I saw only because one was next door to our hotel, was Bunratty's Castle. As castles go, it looked small—not nearly large enough for some swashbuckler to leap from a balustrade, swing through the air on the end of a thirty-foot tapestry, and land on a parapet on the opposite side of a stone-walled great room. But then, aside from the medieval fancies of Hollywood set designers, I had never seen a real castle before.

Durty Nelly's, a centuries-old pub pieced together wing by wing and gable by gable, was a representation of the latter. The small rooms with rough-hewn floors sanded to a smooth finish by millions of tipsy Irish feet that had scuffed their way to the bar, one in each room and more upstairs, so that one was never far from a pint or two of stout. It looked authentic enough, as did the old farm tools and medieval weapons hanging on the lumpy plaster walls. There were flagons and shields with coats of arm that hinted at the forgotten glory of the O'Brien or the McGaedy clan.

Campbell and I drank a glass of lager to wash down the stale taste in our mouths and went back to the hotel, where I interpreted a more pleasant meaning when the pretty desk clerk confirmed my wake-up call. Using her native colloquium—which was quite different from the American—she cheerily promised to "knock me up at one A.M."

The radio was ominously silent as we crossed over Europe. The English, and later the French, controllers barked infrequent directions to the lone private or military aircraft until dawn came over southern France and, with it, the airlines cranking up for a new day's business. The cloud-covered landscape of central Europe looked like a field of moldy clay, glistening in the pink-yellow light of sunrise, lumpy and rolling in cumulus puffs that broke on France's Alpine peaks and disappeared as we approached the Mediterranean.

The Riviera passed under our wings as we began the descent to Genoa in a cloudless sky. Due to the smut-brown smog of industrial pollution, the airport was invisible until I was only three or four miles away. Then it appeared in my windshield as if perched on a promontory on a narrow strip of rocky ground between the mirror-smooth Mediterranean to the south and the mountains to the north. It was just as Flournoy warned me about back in Lock Haven.

"Genoa's not bad if it's just hazy, like it usually is," he had said. "But if they've got low clouds and you have to shoot the approach, try not to get off course to the north." While offering this advice, he made the sign of the cross to indicate the vertical and horizontal needles of the aircraft's instrument landing system, and also, I suspect, to suggest prayer to help me stay precisely on course throughout the approach.

Whatever luck I had had in stumbling into ferrying planes in the first place was magnified by the fortune that allowed me to triumph in an endeavor that could quickly take one's life with neither warning nor reason.

First the flight controls freezing *after* there was enough warm air below me to melt the ice before hitting the water, and now not having to fly an instrument approach with mountains so close to the runway. It was difficult not to believe such providence to be the workings of a higher power. If there were gods controlling our fate and if you had to cross the ocean with a briefing by someone called "Doomsday," it stood to reason that prayer probably wouldn't hurt.

7

Land of the Midnight Sun

A retired Air Force transport pilot who used to fly heavy, four-engine DC-6s and Boeing Stratocruisers once told me that the ocean was no place for single-engine airplanes and that when he was over water he wanted to be able to look out the window and see engine after engine pulling the airplane through the sky.

It was a wise and prudent statement but only a pipe dream as far as I was concerned. Most of the airplanes Globe Aero ferried were single-engines, which is why I couldn't understand my good fortune in being assigned four twins in a row. Especially since that was the total number of airplanes I had taken. It wasn't that an airplane with two engines was better than an airplane with only one, but when you're a thousand miles from land and an engine quits, there's a distinct advantage in having a second engine on the airplane.

Sooner or later Moody would have to put me in a single-engine or else the other pilots would start complaining about the new guy who was flying nothing but twins. Any antagonism this might cause wouldn't be from envy as much as self-preservation: the more twins I flew meant more singles everyone else would have to fly. The time to pay my dues, as it were, came on my next trip, when Moody assigned me a small, single-engine Cessna for Sweden.

The airplane was parked on the ramp in the same spot it had been a week earlier, after the extra fuel tanks were installed, and it was rolled outside to wait for a pilot. But as each pilot returned from overseas and saw the airplane, they suddenly remembered a previous commitment, which caused them to turn down the trip. When I went outside to look the airplane over, I could understand why.

It was small, all right, and crammed with the usual extra ferry tanks and fuel lines. The worst thing about the airplane was that it didn't have any permanent radios—not even a single VHF comm. Instead, there was an entire set of temporary radios mounted on top of the fuel tank to the right of the pilot's seat.

We didn't get many airplanes without radios, but even one was too many. They were a nuisance, not to mention unsafe—with electrical wires close to fuel tanks—just so the buyer could save a few dollars by installing the radios himself. For the pilot who had to remove the radios after delivering the airplane and lug them back to Lock Haven, it was a curse. Hoping to make that part of the delivery as painless as possible, I asked Probst how I should go about accomplishing that task.

"It's good that you came and asked me," Probst said, "because I'm tired of getting radios back with knobs broken off, wires ripped loose, and looking like they've been used for a hockey puck. What do you guys do, put 'em in with checked baggage? I know everyone's in a hurry to detank and pack everything. And once the radios have done their job, being careful with them is probably a low priority. But if I keep getting banged-up radios back, guess whose airplane I'm going to save them for."

"I can only imagine," I said.

"They may not be in the best of shape; they're scratchy, reception's lousy, and they've probably been in more airplanes than you've had hot Sunday dinners. But Walt thinks he can get a few more years' use out of them. So the best thing you can do when you detank is allow a little extra time to disconnect and pack the radios."

I was careful not to jostle any of the wires or radios as I began loading my luggage. There was barely enough room for a suitcase, let alone my survival gear and flight bag. Yet, somehow I fit everything in, cramming a bag here and a bag there as if I was packing sardines.

The airplane used two-thirds of the short runway on takeoff—and that was with only five gallons of fuel in each ferry tank. It would be much worse when they were completely full. With any luck I would have to do that only once, in Gander, where the runway is four times longer than Lock Haven's. That was still two or three days away. Today, I was going only as far as Bangor, Maine, and then on to Moncton, New Brunswick, tomorrow for my appointment with the Canadian Ministry of Transport to get my single-engine waiver. A waiver authorizes a pilot to depart Canadian airspace in single-engine aircraft for flights over the ocean and is issued after the pilot has successfully completed oral and written tests on North Atlantic weather, navigation, flight planning, and ditching procedures. I had already studied for the test, so when I landed in Bangor I spent what was left of the day getting to know the city.

Bangor is a regular stopover for most of the transatlantic ferry pilots who come here to export their aircraft before leaving the country. The third largest city in Maine, it boasts a large, underused international airport that is an ideal place for ferry pilots, overseas charter airlines, and military flights. Before the airplane was even a glint in the Wright Brothers' eyes, however, Bangor was a booming lumber town. The old timers, spinning tales handed down from their grandparents, tell of a time when a never-ending procession of logs floated down the Penobscot River on their way to the sawmills. Like gold-rush towns of the Old West, the main street of what was then known as "The Queen City of the North" was lined with brothels and saloons.

Those bawdy establishments, along with the colorful era that gave them birth, now belong to Bangor's past, hinted at by the twenty-foot-tall statue of Paul Bunyan standing watch across the street from my hotel. Bangor is now a small New England town like many others, except for the tourists and flight crews, whose German and British accents seem at odds with the down-home inflection of the locals. While people still flock to Bangor for the trees, instead of cutting them down, they come to gaze at the rich autumnal hues of their foliage. They also come for the lobster.

As I had been told numerous times back in Lock Haven, if you wanted a good lobster dinner in Bangor there was only one place to go—The Pilot's Grill. I didn't know if it was because the lobster was that good or that it was a superstitious ritual, but as far as most of Globe Aero's pilots were concerned, an Atlantic crossing shouldn't even be contemplated until one availed themselves of this succulent crustacean. Not seeing the connection between luck and food, I opted instead for the Chinese restaurant across the street from the hotel, where, after the meal, I cracked open my fortune cookie to find it empty.

What did this mean? I wondered. If taken literally, it was obvious—I had no future. Except this wasn't ancient China, it was the twentieth century. You couldn't put a curse on people any more than you could see your future in a cookie. Some bored Chinese cookie maker neglected to put a fortune in a few cookies. It was a simple mistake with no mystical glimpse of one's fate. Yet, I still felt uneasy.

The training I had received from Waldman and Campbell paid off the next morning when, four hours after meeting with the MOT inspector in Moncton, I walked out with a new single-engine waiver in my pocket. I don't usually do well on tests, so I was relieved now that this was out of the way. The elation lasted until I called the weather office in Gander.

"Now, you don't really want to come up here today, eh," a jolly voice that identified itself as Clayton Jeans echoed over the wires.

"I was thinking about it."

"Oh, you know we've got less than half a mile visibility in fog and . . . let's see . . . yep, Torbay's got the same. Typical spring weather, eh. We've got a low moving up from the south and just about every place is down. We haven't seen the sun for three days now, and I suspect it'll be like this for a few days more at least."

"I guess with that low out there I can forget about going to Shannon?"

"Well, certainly not great-circle. But . . . wait just a minute. Let me look at something else." I could hear him shuffling papers before returning to the phone.

"Okay, got it all right here now. If you want to cross tonight, the best thing to do is go up to Goose Bay. They're wide open and that low won't be affecting their weather. Then you might want to think about going the northern route to Reykjavik."

"I don't know, Clayton. I've never flown that route before," I told him, hesitant about deviating from the normal route on only my second trip across the Atlantic.

"Well, my son," he confided, probably sensing my hesitation. "There's a high over Greenland, so that shouldn't be any problem, eh. The only thing you might have to worry about is if the low picks up speed. Then there'd be some clouds over Iceland. But even then, the worst you'll have is a two-thousand-foot freezing level."

I wasn't entirely comfortable with the idea of flying over the ocean in a single-engine airplane to begin with; now I had to fly a route I hadn't flown before. But Clayton was reassuring, and, because I was ferrying airplanes because I didn't like routine, I took off and headed north to Labrador.

Goose Bay is three hundred miles north of Newfoundland and was used by ferry flights when Gander's weather was down or when we didn't have enough fuel to go straight across. It was in the northern part of Labrador, a vast expanse of Atlantic Canada where small towns with names like Mont-Joli, Sept-Isle, and Lake Eon are separated by an endless horizon of tundra. Like the outback in Australia, it was a lonely flight over a barren landscape until I reached Goose Bay.

The sign on the small terminal building said WELCOME TO HAPPY VALLEY. Strange name, I thought, for such a bleak civil-military airport so close to the end of the North American continent. When I was in the Air Force, a fellow GI who had been stationed here once told me he let his paychecks accumulate uncashed for several months because there was nothing to spend them on. I suspect it was called Happy Valley for the same reason the Vikings named Greenland—to lure settlers with a deceptively welcome-sounding name.

In any case, I wouldn't be here long enough to find out. I planned to refuel, check the weather, and leave for Reykjavik as soon as I could. When I went into base ops, however, the Royal Canadian Air Force sergeant was briefing another pilot on weather over the ocean.

"Where're you going?" I asked, thinking he too probably had to take the northern route and we could fly as far as Reykjavik together.

"I will go to Narsarsuaq tomorrow, then Reykjavik, and finally to Sweden," he said, in a slight Scandinavian accent. "And you?"

"That's where I'm going. To Umea, up in the northern part of the country. Is that anywhere near where you're going?"

"Yes. Now this is a coincidence. I am going to Ornskoldsvik. Umea is only forty miles from there."

It wasn't just a coincidence; it was kind of spooky. "Now, what are the odds of that," I said, "running into someone in Goose Bay, of all places, and find out we're not only going to the same country, but within forty miles of each other."

"We can fly together, yes?" he suggested. "It is a long way to Sweden."

"I was thinking the same thing. But I was planning on flying direct to Reykjavik. I don't know if I want to make an extra stop. Do you have to land in Greenland?"

"I do not have enough fuel to reach Iceland so, yes, I must stop in Greenland. But why do you not stop there also? It is only seven hundred miles from here, and we can do that all in daylight. If you go direct to Reykjavik, you will pass over Prinz Christiansund on the southern tip of Greenland. Narsarsuaq is on the west coast only sixty miles north. It is not that far out of the way, and, the briefer says the weather should be good."

"I guess I could split the flight into two short legs and do them both in daylight, rather than one long leg, part of which would have to be at night," I said. What I neglected to say was that I still didn't feel comfortable with the airplane or the route and would rather fly in daylight. More important, I wanted the company. Ferry pilots enjoyed a sort of instant camaraderie and often went miles out of their way just to have someone else to combat the loneliness. Friendships were made fast and trust was shared unequivocally. I didn't find out until we were in the taxi driving to the hotel that the pilot's name was Karl Nilsson, a Swede living in Mexico and flying home for vacation.

We spent the night at the "Goose Hilton," an old army barracks that had been turned into a hotel, and got an early start the next morning. It was just after sunrise, and the sky was a rich blue without a cloud in sight. The forecast for Narsarsuaq looked good, and, considering that the distance was so short, I didn't expect navigation to be a problem. Two of the major elements of a successful crossing, weather and navigation, were in my favor. That left only the third—mechanical problems—specifically, the engine. Without warning it could blow a cylinder and seize while I was a long way from land. I tried to reassure myself with the argument that I was only asking a brand new engine to keep running until I got to Sweden—twenty, maybe twenty-five hours more. It didn't seem too much to ask.

As I taxied to the runway, I checked pressures, temperatures, and rpm, and I found nothing that was not within limits. Satisfied that everything was as it should be, I moved the throttle to full power. As speed slowly increased I felt the aircraft becoming more responsive to the controls. When it reached eighty knots I eased back on the control wheel.

A shudder went through the airplane as it broke free of the runway friction and seemed to leap forward. It was only a relative perception—since my airspeed increased only a few knots and, because of the extra fuel, the aircraft was barely climbing. I hoarded each one-hundred-foot increase in altitude like a miser, knowing that if the engine were to quit now, as heavy as the airplane was, I wouldn't have enough altitude to glide back to the runway, but would crash-land on the tundra.

For the first few hours I listened to the engine for even the slightest change in pitch or skip in its revolutions. My eyes were glued to the engine gauges, staring at them so intently I could have bored a hole right through the glass. Nothing. Throughout the flight, the dials and needles were so rock-steady that they might have been welded in place. Nevertheless, I was glad to see the snow-capped peaks of Greenland come into view and grow larger as I approached the glacier.

"I think I see the fjord," Nilsson announced over the radio. His airplane was a few knots faster than mine and, as such, he was my trusty scout on what for us was a pioneering flight.

"Are you sure it's the right one?"

"Well . . . I think so. But I will know soon."

There was no published instrument approach for Narsarsuaq, only a visual letdown chart that depicted three fjords, two of which were deadends that terminated in a wall of rock and one, forty miles long, that opened into a large clearing where the airport was located.

A few minutes later Nilsson came back with the news that he was in the correct fjord. "It is clear, so you can either fly down the fjord, also, or stay at altitude until you are over the airport and then spiral down. But I would hate to have to come in here underneath a low overcast, because I do not think you will find it so easy."

Realizing that someday I might have to do just that, I figured it would be a good idea to see what the approach looked like when the weather was clear. I began a descent down to a thousand feet and looked for the fjord that led to the airport. As I crossed over the Simutaq radio beacon, I saw the entrances to the three fjords that wound into the glacier. Julianahab Radio cleared me for the approach, and I dropped down to eight hundred feet as I entered the southernmost fjord.

Any apprehensions I had had about coming here were swept away as the majesty of this magnificent scene unfolded around me. It was eerie to look up on either side at the shear walls of rock towering above me as I zigzagged down the fjord. I was twenty miles into the fjord when I saw a sunken ship with its hull sticking half out of the water. Years ago the ship had sunk and was left, I supposed, as a landmark to aid navigators. This might be okay for ships that ventured into the wrong fjord and could stop or turn around. Airplanes, however, were not as easily maneuvered; especially if the clouds were low and the pilot was scud-running a few hundred feet above the water. Because twenty miles into any one of the fjords, if the ship was not sighted, there wasn't room to turn around and fly back out.

I banked the airplane around a mountain peak jutting up out of the water and entered a large valley. The airport was just to my right, neatly tucked into the southern corner with a three-thousand-foot vertical wall of rock a quarter of a mile to the side of the runway. It would be quite a surprise for anyone attempting to navigate through the fjord in low visibility. I cut the power, put the airplane into a rapid descent, and rolled out over the end of the runway.

There was little activity on the airfield. Nilsson was refueling his airplane, and a Greenland Air helicopter was parked across from him. No other aircraft were in sight. Narsarsuaq might have been a busy place during World War Two, when it was known as Bluie West One and used by the Allies to ferry bombers and supplies to Europe, but now it was a quiet, little settlement. Five or six structures made up the entire village: a few hangars, the airport terminal, and the Arctic Hotel, which had the distinction of having Narsarsuaq's only bar. As such, it was the center of social activity for this motley settlement of Inuit Eskimos, Danish-Norwegian adventurers, and the handful of Greenland Air pilots, who were mostly expatriates from northern European countries.

Later that evening, after an expensive dinner in the hotel restaurant, Nilsson and I were having a drink at the bar when we were joined by the rotund Dane who had refueled our aircraft earlier. Everyone seemed in a holiday mood, chatting in various mixtures of Danish, Inuit, and broken English, as they washed down their meal with lots of Danish beer.

"You choose a good time to come here," said the Danish fueler, and then took a long swig from his mug. "Today, the first ship this year comes to Narsarsuaq."

"That seems a good enough reason to celebrate," I commented.

"Ja, it is very good reason," he said, draining his mug and setting it down on the bar. "The ship was carrying a new supply of beer."

"You mean you have to wait until summer to get supplies?" I asked. "What do you do in the winter when ships can't come down the fjord and you run out of liquor?"

"That does not happen often, since we have a large supply we keep in the cave your United States Air Force used to store ammunition in during the war. But sometimes, if the winter is long, we do run low."

"What do you mean *if* the winter is long?" asked Nilsson. "Winter is always long on Greenland. Your supply must always be low."

"Then we fly to the capital in Godthab to get more."

"You mean Thule," I corrected him.

"Thule? Why would we go all the way to Thule?"

"Because that's where the capital is," I said, figuring I should correct this Dane. After all, he only lived here.

"No, you are wrong, my friend," the Dane informed me. "The capital is Godthab."

World traveler and expert in geography that I was, I knew of the big U.S. airbase in Thule and figured that had to be where the capital was. I was certain, and I was sticking with my answer. The Dane, now joined by the bartender, was equally adamant that the capital was in Godthab. We argued this point back and forth until Nilsson—who, as a European, was more aware of Greenland's status as a territory of Denmark—informed us, "All of you are wrong. The capital of Greenland is in Copenhagen."

I was flying in tightening circles, trying to gain altitude to escape from the boxed canyon, but the airplane wouldn't climb. I pulled harder on the control wheel, attempting to get the nose up. It felt as if it weighed a ton, and the more I pulled the faster it sank. I kept circling as I watched the sides of the canyon get closer. Too close! The trees and jagged edges of the canyon floor were rushing up at me. I pulled back on the control wheel with everything I had but was still being sucked down into the darkening abyss. The airplane began skimming the tops of the trees—when everything went dark.

I had been having this same dream ever since I learned to fly, and I had no idea what it meant. A month, maybe two, would go by before I put it out of my mind. Then I would have it again and awaken terrified, not sure what was dream and what was reality. Then feelings of helplessness and imminent death haunted me until I realized it was only a dream. Was it common for pilots to have nightmares of crashing, or was it a premonition? Was it symbolic of some deep-rooted fear of being out of control, or was it a precise warning to stay out of boxed canyons?

I probably would have forgotten about it, as I usually did, except that the boxed canyon in my dream looked a lot like the clearing that Narsarsuaq was in—the same clearing that I had yet to fly out of. I kept thinking about the dream during breakfast, my mood turning gloomier each minute, even as I tried to convince myself it was just a coincidence. The Danish control tower operator didn't help when he briefed Nilsson and me on the standard procedure for departing the airport.

"You must be at nine thousand feet to fly over the ice cap. So after takeoff you have to circle and climb to seven thousand feet before turning on course," he said, in a thick Scandinavian accent that made me think of an Ingmar Bergman movie where someone was playing chess with the devil.

That did it! Unknowingly, the briefer had just described exactly the setting of my dream. Over the years I hadn't given it much thought, but now I saw it as a prophecy. I didn't think I was superstitious, yet I couldn't help seeing all the little things that had been happening on this trip as signs: my first single-engine crossing, the full set of temporary radios, and having to fly a route I was unfamiliar with. Then there was the chance meeting with Nilsson, whose destination was so close to mine. The stop at Narsarsuaq, magnificent yet eerie, was by no means a normal stopover in the North Atlantic. And now the dream.

Each sign alone was meaningless. Even together, although disturbing, they were probably no more than coincidence. Then I remembered the Chinese restaurant in Bangor . . . and the fortune cookie. I knew now that the empty fortune cookie was no mistake. It was a warning.

"You do not look good," said Nilsson, as we walked to the aircraft. "Too much scotch?"

I couldn't tell him the truth because it would've sounded ridiculous. Instead, I stalled. I looked for any excuse to delay the inevitable. I could pretend I had a mechanical problem, but that would only postpone it. Or I could back out completely, call Moody, and tell him . . . what . . . that I couldn't finish the trip because of some vague premonition based on dreams and fortune cookies. I didn't want to do that either. It was almost noon, and I couldn't procrastinate any longer. The only thing I could do was see the trip through to its end, whatever that might be.

Nothing unusual happened during takeoff, nor when I began the circling climb. The clearing was large enough to make lazy circles in and seemed to fit the architecture of my dream. The bright blue sky was cloudless. I thought it was dark in the dream, but I wasn't sure. Anyway, that could have been symbolic. But where were the trees? I know there were trees in the dream, because I remember them scraping the bottom of the airplane. Maybe the dream wasn't a prophecy after all, and my dread was all for nothing. But there was still the fortune cookie.

The airplane spiraled upward slowly—three thousand . . . four thousand . . . five thousand feet. At any minute I expected the engine to cough and sputter to a stop. But it didn't. It continued climbing through seven thousand feet as I rolled onto an easterly heading. Below, the airport and buildings looked like the village of a model railroad laid out in minute detail. In front, the glaciered top of the ice cap was moving closer, agonizingly slow, until it slipped beneath me and I was clear of the valley.

Although I had escaped the warning of my dream, my eyes now took in the unearthly surrealism and timelessness of Greenland's ice cap, vast and boundless, as it disappeared over the northern horizon. In the crisp, arctic air everything stood out so clearly that windswept peaks off in the distance were as sharp and focused as fjords at the edge of the glacier a mere forty miles or so in front. A pristine stillness hung over the frozen landscape, and I imagined, but did not see, swirling drifts of snow that covered all traces of life.

Everything was deathly still. The vast landscape of virgin tundra quietly stretched out to distant mountains, the instruments in my aircraft were frozen in place, and the engine hummed steadily. I didn't move for fear of upsetting this delicate balance between order and chaos. The ice cap was hypnotic, looking far more forbidding than the ocean ever did.

My journey into this serene, unspoiled world lasted no more than a few hours before I was over water on course for Iceland. Nilsson, twenty minutes ahead of me, saw it first and, like a trusted Indian scout of the Old West, informed me of what I would soon be flying into.

"I can see the clouds from here, and it does not look good," Nilsson said. "Reykjavik is reporting seven-hundred-foot overcast and four kilometers visibility, and they say it is getting worse."

Thirty minutes later, I saw the clouds Nilsson warned me about, appearing at first as no more than a thin line drawn between sky and ocean. As I flew farther east, the line grew thicker, crooked, and lumpy with a darkening underbelly. In the distance, almost dead on course, was a mountain of cumulus that was turning the sky black. It was the low-pressure area Jeans had spoke of earlier. The sea below me had given in to an invasion of small, cotton-candy puffs that expanded and grew until they formed a solid layer.

Nilsson had just landed when he called over the radio with the latest weather at Reykjavik. "The ceiling is dropping very fast. You should hurry if you want to get in, because it was down to three hundred feet and two kilometers when I broke out. And watch out for the church steeple. It is only a half mile from the runway."

Of course there would be a church steeple—or a tower—I thought. It wouldn't be a normal airport if there weren't. Hurrying to get down, however, was the one thing I couldn't do. The airplane was already at full power, and there wasn't anything I could do to make it go faster. I was twenty-five minutes from the airport. As fast as the ceiling was coming down, the best I could hope for was that it would be at minimums when I arrived.

With a 244-foot ceiling and three-quarters of a mile visibility, I would have to be precisely on course throughout the approach. This would not be a problem if the airplane had the right radios.

The instrument approach at Reykjavik was an ILS—short for instrument landing system—which emitted two electronic beams from a ground station at the touchdown end of the runway. One was a vertical beam called a localizer and the other was a horizontal beam called a glide slope. Both beams were represented in the airplane on an instrument called a glide slope indicator by vertical and horizontal needles that moved away from their respective centers of the instrument if the airplane drifted too far right or left or too high or low. Like the crosshairs on a gunsight, all I had to do was keep both needles lined up in the center of the instrument to be led down to the centerline of the runway. The Cessna I was flying was equipped with the indicator needed for this approach, except it wasn't where it should have been.

When flying in clouds solely by reference to instruments everything should be directly in front of the pilot to keep head movement down to a minimum. Navigation instruments that show the aircraft's position over the ground should be next to gyro instruments that depict heading and whether or not the wings are level. Clustered around these important instruments are others that depict airspeed, altitude, vertical speed, and rate of bank.

In my Cessna the glide slope indicator was installed along with the rest of the temporary radios on top of the fuel tank where the copilot's seat used to be. This meant I would have to do exactly what I shouldn't do. I would have to watch the gyro instruments in front and then turn my head quickly and repeatedly to the side to see if I was still lined up on the ILS. Although minimal, any movement was probably too much because it wouldn't take long for the nauseating symptoms of vertigo to cause me to lose my sense of balance and become disoriented.

Then how long before I began to disbelieve what the flight instruments were trying to tell me. And finally, attempting to keep the aircraft level by the pressure I felt in the seat of my pants. That would be fatal. It had been happening since man first challenged the clouds in instrument flight and usually ended with the pilot becoming so disoriented with up, down, right, and left that the airplane fell off into a spin and corkscrewed into the ground.

Even if my instrument-flying skills were good enough that I could keep the airplane from entering a spin, I could easily drift off course and run into one of those hills or the church steeple, which would be worse. About the only thing I could do was stay out of the clouds as long as possible. I knew from training that the longer I flew on instruments, the more susceptible I was to vertigo. Fifteen minutes had passed since Nilsson landed, so I called Reykjavik to see if there was any change in the weather.

"Sierra-Gulf-India, from Reykjavik Approach," the Icelandic controller answered in clear, precise English. "We are showing eight hundred meters visibility, two-hundred-foot ceiling, and barometric pressure at nine-nine-eight millibars and falling rapidly."

I made a quick mental conversion of roughly three feet per meter and came up with twenty-four hundred feet—about half a mile. The weather was dropping too fast. I still had thirteen minutes to fly to the airport, and the ceiling was already below the published minimums, which meant legally I couldn't even start the approach.

I waited for the controller to ask what my intentions were, hinting I wouldn't be cleared for the approach and should proceed to another airport. But he never did. Apparently, since low clouds and fog were not that uncommon in Iceland, they took a more lenient attitude toward regulations and left it up to the pilot to determine his own level of skill and bravery. It was a level I did not want to test. I asked the controller for the latest weather at Akureyri, one hundred sixty miles away, on the northern coast of Iceland.

"They are reporting just above minimums at three thousand meters and eight hundred feet," he answered. "But they are dropping rapidly. I do not think you can get there in time."

Although the weather was better, Akureyri was inside a fjord with five-thousand-foot peaks on either side. I didn't want to experiment with that type of terrain even if I could get there before it went below minimums. "How about Keflavik?" I asked.

"Unfortunately, Keflavik just went zero-zero in fog and is now closed."

That explained the controller's leniency: the entire island was socked in. It looked as if the closest airport I could get into was back at Narsarsuaq or

Kulusuk on the east coast of Greenland. I had enough fuel to stay in the air three more hours, but that wasn't enough to reach either airport. The only place I could land was Reykjavik, and it would have to be soon, before it, too, went zero-zero.

I crossed over the outer marker and lowered the nose to intercept the glide slope, keeping the engine at full power and letting my airspeed increase. Milky gray clouds immediately enveloped me, completely blocking the horizon. A strong crosswind pushed the airplane to the left. I gently stepped on the right rudder pedal to stop the drift and get the needle back in the center, because I didn't want to overcorrect and start drifting in the opposite direction. Airspeed was one hundred forty knots, and I was coming down fast. The ILS needles were responding to the slightest pressure on the controls. I was moving my head back and forth between instruments, each time robbing myself of seconds that allowed the airplane to drift off its heading or porpoise up and down on the glide slope.

I passed the five mile fix; two and a half minutes to go. I was at one thousand feet. Reykjavik's elevation was forty-five feet above sea level. I had to remember to subtract that from what my altimeter indicated.

"Sierra-Gulf-India, we just had a commuter airliner land five minutes ago, and he reported six hundred meters visibility and two-hundred-foot ceiling. This is the absolute minimum for landing," the controller warned.

I presumed that was the controller's way of telling me I would get this one chance only, and if I didn't land, he could not clear me for another approach. Without enough fuel to make it back to Greenland, my only alternative would be to put the aircraft into the water. I was not going to disappoint him . . . or myself.

Five hundred feet. The airplane was dropping below the glide slope. The beams were narrowing as I got closer to the ground, and the ILS was becoming more sensitive. I swung my head to the gyro in front. The airplane was in a twenty-degree bank. I shook my head to stop the vertigo from getting worse, eased back on the control wheel, and leveled the wings.

Two hundred seventy feet. The altimeter was unwinding quickly. I still had another forty feet before I would break out of the clouds, *if* the ceiling didn't come down any more. What if it did? How low should I descend? How long before I drove straight into the ground?

I had to get down before the fog rolled in and engulfed the airport. If my airspeed got too fast, however, I would overshoot the short runway. I had to slow down. I pulled the throttle back to half power. The localizer needle began swinging to the left. I put pressure on the left rudder pedal, though not too much, because I was below the church steeple.

At 170 feet above ground I saw the approach lights, barely visible in the corner of the windshield. Seeing nothing else but vague shades of gray, I kept my eyes glued to the lights piercing through the fog and leading me to the runway. I closed the throttle, lowered the flaps to full down, kicked in the left rudder, and swung the control wheel in the opposite direction. Cross controlling at this speed was an almost sure way to go into a spin, but it was the only way I could get rid of airspeed and altitude quickly.

The airplane dropped out of the sky with all the grace of a grand piano. I passed over the threshold lights, rolled out of the slip a few feet above the runway, touched down, and brought the airplane to a stop. I could see no farther than a few hundred yards in any direction, and even that was only a half dozen or so runway lights that disappeared into the gray emptiness. I turned off the runway onto what I hoped was the parking apron and shut the aircraft down in front of a large building looming out of the fog.

"Welcome to Reykjavik," Nilsson said, as I entered the door marked FLIGHT SERVICES. "I did not think you were going to get in."

"Ja. The weather gets much worse. Just after you landed, the airport closed," said the ruddy Icelander behind the counter who introduced himself as Sveinn Björnsson, handling agent for all private aircraft that land at Reykjavik. "It is your first time in Iceland?"

"Yes, and I hope the weather's not always this bad," I said.

"This is not bad. In winter, when we have only a few hours of daylight and storms pass through here one after another, *then* it is bad," said Björnsson, with a knowing snicker. "This fog, though, it will not last—because it stays light now. Today is the longest day of the year."

"Of course!" Nilsson suddenly remembered. "June twenty-first. The summer solstice—when you can see the midnight sun."

Pagan ceremonies with elves and trolls and fair-haired sun worshipers dancing in a field of dandelions passed quickly through my head.

"In the north, maybe, but not in Reykjavik," said Björnsson, kindly correcting any mythical misconception I might have had. "It stays light out, but you cannot see the sun, and so I do not think you will find many people outside tonight." Besides arranging for weather briefings and helping with flight planning, Björnsson also acted as a concierge, making hotel reservations, suggesting restaurants, and dispensing advice about the island and its people.

It was close to midnight, yet still light out when Nilsson and I left the restaurant. The fog had lifted, leaving only a high, thin overcast covering the city. Björnsson had been right about the scarcity of people, for the streets were deserted. It felt as if we were in a ghost town. Walking along the narrow streets of this colorful settlement, one could easily imagine the specters hiding in narrow alleys and underneath stairwells are the elves and trolls of old pagan Viking legends.

The low had passed to the northeast and gave us no more trouble when we left the following evening for Sweden. Even the cold front off the coast of Norway was weaker than forecast. Not a burble of rough air and only a few strands of wispy stratus that were over quickly. I felt foolish for having worried all the time I was over water, yet relieved as each minute brought me two miles closer to the end of my journey. Just a little farther and, even if the engine were to quit, I knew I could get the airplane down and walk away from it.

It didn't look as if that was going to happen, because the engine that I had been worrying about ever since I left Goose Bay never so much as sputtered in protest. It continued its faithful labor right up to when I landed at Ornskoldsvik and let it wind down for another night's rest. Tomorrow I would make the short hop to Umea, turn the airplane over to the customer, and be done with it.

As for the dream, who could say whether its nebulous meaning was superstition, anxiety over the flight, or a portent to be heeded. I didn't know enough about the supernatural or the science of dreams to answer this question. I still felt that it was trying to tell me something—maybe not to pay so much attention to signs and omens or that flying a single-engine airplane over an unfamiliar route was no more dangerous than flying from Gander to

Shannon. Maybe it was cautioning me to be careful not only with airplanes, but with life itself. Or possibly it was yet another warning of a boxed canyon somewhere that I would not escape.

When I had time to think about it, I wasn't sure if the dream was a recurring one or not. Maybe it was just so powerful and affected me so strongly that I only believed I had had it before. Or could I have dreamed that I had dreamt it before? These were questions for which I would never know the answer. I did know one thing—the nightmare was absent from my sleep in Reykjavik. Whether I had beaten fate or had enacted and overcome some primal fear, I have no idea. Whether I had eluded the demons of my sleep for the first or the hundredth time, I never had the dream again.

8

A Ship without a Rudder

Over the next few months, every time I went overseas there was only one thing people wanted to talk about: Watergate. The black cloud that hung over President Richard Nixon during the summer of 1974 eventually gave way to the golden hues of autumn and, as if spurred by the change in seasons, interest in the political scandal waned. The growth of general aviation that had begun three years earlier, however, showed no signs of letting up. Piper, Beech, and Cessna were introducing entire fleets of new models and churning out airplanes in record numbers. Numbers that in previous decades had been no more than a few hundred now measured in the thousands. There was enough business for anyone who hung out a shingle and called themselves a ferry company.

Although there were a handful of one- or two-man operations, most of the aircraft going overseas were ferried by one of the three companies that had aligned themselves with the manufacturers. There was Flo-Air in Wichita, Kansas, whose pilots flew only Cessnas, and because of their higher pay, were considered the elite of our profession. Air Facilities in Oklahoma City, other than the few Beech aircraft we ferried to Australia, delivered most of that manufacturer's airplanes. As for Globe Aero, just across the runway from the main Piper factory, our relationship was the most lucrative of all. Of the three big makers of small airplanes, Piper was by far the largest with three plants in full production and a growing network of overseas distributors that bought up airplanes as fast as they came off the production line.

Our economic prosperity, for the near future at least, was assured. Every time I came off a trip there were more aircraft parked outside our hangar, and there were more pilots' names on the board beside Moody's desk. They were young, old, ex-military or ex-airline, some just beginning their careers in aviation and others with thousands of hours of heavy jet time in their logbooks. They had come to the right place at the right time, because Globe Aero had so many airplanes going to so many parts of the world, rarely a day went by that we didn't spend some portion of it in the air.

Unlike airline or charter pilots, who were limited by regulations to no more than one hundred flying hours per month, we had no such restriction. Ferrying airplanes, an enterprise that did not produce revenue per flight, was considered private flying, and under more lenient rules we could fly as much and as often as we wanted. One hundred fifty hours a month was not uncommon, and there were many who flirted with two hundred, mindless of the price for such globe-trotting adventure. We froze in wintry snow-encased villages in the Arctic Circle one week and sweated in the heat and humidity of sultry tropical islands the next week. We crossed time zones at such a dizzying pace that our internal clock was in a constant state of war with local sleeping habits. We got punchy after flying for several hours in the oxygen-thin altitudes above fifteen thousand feet. And we often had trouble coming up with an answer when an immigration officer at JFK Airport or some other port of entry asked us what country we just came from.

We ignorantly tolerated the toll this abuse was taking with a nonchalant shrug that payment would be claimed, if at all, at some imagined time in the distant future. For the self-imagined prestige of being in a profession where our total number of colleagues throughout the world was counted in two-digit numbers far eclipsed any thoughts we might give to rest. Instead, we used to race to see who could fly the fastest trip or deliver the most aircraft in a year, a month, or even a week.

"We were always racing for some reason or other," Nils Mantzoros told me, while waiting for our next trip assignment. Mantzoros was the fourth pilot Moody had hired back in the early years before the flood, when airplanes were scarce and pilots were lucky to fly one or two trips a month. Although this was hardly the best time to have been ferrying airplanes, the few pilots who flew for Moody back then looked on those days fondly.

"It started getting busy after the flood," Mantzoros reminisced. "Piper had a backlog of airplanes, so we had to start hustling. But what really got things going was when they came out with the Seneca [a six-passenger twin]. It was just the shot in the arm general aviation needed."

Mantzoros spoke with a quiet authority and knowledge uncharacteristic of his scant years or size and was respected by his fellow pilots, young and old alike. He was barely old enough to drink and still traveled on a student ID card when he started flying for Moody, but he took to ferrying airplanes as if he were born to it. When the other pilots at Globe Aero were flying to Europe and back in two or three days, Mantzoros took a Seneca to Dakar, Senegal, and was back in Lock Haven thirty-six hours later.

"I guess I just got lucky, because I wasn't even trying to run a fast trip. It just worked out that way," Mantzoros said. "I got to Gander and there was a good tailwind, so I left right away, flew all night, and landed in Dakar in the morning; something like two hours early. Still, it looked like it'd be a normal trip until I saw Pan Am taxiing in. I figured it had to be the New York flight, so I called the captain and told him what I was doing there and that I'd really like to get a ride back with him but I had to drop this airplane off first. 'Well, okay,' he said. 'I'll drag my feet a little. See if you can make it.' "

"You must've really had to hustle. It takes me an hour to get all the tanks out and pack everything. I don't think even the most senior Pan Am captain could drag his feet that long."

"That was the easy part. I'd already been detanking it in flight."

"Isn't that kind of dangerous?" I asked, knowing full well that it was.

"Oh, it can get pretty tense if you've just disconnected a fuel line and you've got your thumb keeping the fuel from squirting all over the place, and then you have to drop everything to pull the airplane out of a dive. But I had an autopilot so it was pretty easy. All I had to do when I landed was hand the airplane over to the customer and hustle over to the terminal. I got on the airplane just before they started the engines and was back in Lock Haven that night."

"I bet Walt was surprised to see you back so soon."

"He didn't think I'd left. He kept wanting to know when I was going to leave," Mantzoros said, snickering as he recalled the surprise he had caused among Moody and the other pilots. After that, he became known as Tailwind Mantzoros, because no matter where he went, he was blessed with tailwinds that catapulted him along faster and farther than anyone else. He could be flying with the wind directly behind him and if he had to change course, thirty, sixty, even ninety degrees, it didn't matter, because the wind changed right along with him.

Following the tradition set by Mantzoros and the other senior pilots, we newcomers just kind of fell right in line. We flew one trip after another, each one indistinguishable from the last, as our flying took on the pattern of one long, continuous flight: fly to Gander, refuel, and leave right away for Shannon. We would then make an early morning flight across Europe so we could arrive at our destination in time to catch an airline out the same day. We were always rushing through airports, trying to get to Lock Haven the same night, a schedule difficult at times to accomplish.

Lock Haven wasn't the easiest place to get to. It is a small town snuggled in between two ranges of the Allegheny Mountains with an equally small airport that had no airline service. The closest airport that could handle airliners was twenty-five miles away in Williamsport, but it was serviced by only one airline, Allegheny, forerunner of USAir. The problem with flying on Allegheny to Williamsport is that it operated out of the wrong airport. Instead of JFK, where all the international flights arrived, the Allegheny Commuter departed from Newark. Usually, by the time we landed in New York, we were so tired from traveling that all we wanted was to get the trip over with and catch up on some sleep—only now we had to run a marathon. If there were no delays with baggage or customs, we had just enough time to take the helicopter to Newark and catch the last plane to Williamsport. But if we were late and missed the flight, there weren't many options available. We could take a taxi to the Port Authority Terminal in Manhattan and ride the bus to Lock Haven, rent a car and drive for three hours, or check into a hotel and take the morning flight.

There was, however, a fourth option, which Bennett tried, when he rushed up to the departure gate at Newark a few minutes before the flight was scheduled to depart, only to find the doors already closed and the Allegheny BAC 1-11 being pushed back from the concourse. Not one to let this deter him, Bennett handed his ticket to the gate agent.

"I'm sorry, sir," the agent said. "But we've already pushed back."

"Wait a minute. It's early," Bennett said, indignantly. "I've still got three minutes."

"Well, the airplane's already pushed back."

"But it's not even time, and I've got my ticket," Bennett said, waving it in front of the agent's face.

"We're really sorry, sir, but there's nothing we can do now. The flight's already left. You'll have to go to the ticket counter and let them handle this."

There was a line when he got to the counter and so he waited, getting angrier as the minutes dragged by, until he remembered Ernie Kuney's advice: "You can't back down from them. If you ever get into a hassle, just start screaming and hollering, and they'll give in because they don't want the bad publicity."

Fifteen minutes later, Bennett finally reached the counter and explained the situation to the agent.

"I'm sorry about the problem, sir, but there's nothing we can do," the agent said, in a practiced, emotionless tone, as he handed the ticket to Bennett and then looked to the passenger behind him, clearly indicating that as far as he was concerned the conversation was over.

Bennett, however, was not to be deterred so quickly. "Hey wait a minute. I've got a ticket for your flight that left three minutes early. Now either you're gonna put me in a hotel for the night or give me a ticket or do something to get me home," he said sternly and loudly to the agent, adding a threatening, "or I'll make a scene which you aren't gonna like."

Shrugging his shoulders, and in a bored tone as if commenting on a piece of lint he had just brushed off his lapel, the agent responded, "Go ahead."

Bennett was speechless. He stepped out of line, muttering, as he walked to the airport hotel, vowing never to "fly this goddamn airline again." Given Allegheny's dominance in Williamsport, the threat didn't last long. It would turn out even more ironic several years later when Bennett got a job flying for Allegheny Airlines.

"You should've had a copy of the CAB regs with you," Ernie Wheeland said, when Bennett returned to Lock Haven and told us of his unsuccessful run-in with Allegheny.

"Yeah, a big, thick copy of regs. That's all I need to lug around with all the rest of the stuff I have to carry," I said, sarcastically.

"Well, you can always threaten them, but you see what that accomplishes. If you want to make things easy, you should make room. As much as we fly on the airlines it helps to know what you're entitled to if you miss a connection because of the airline's fault. They'll always try to pull a fast one on you, because if they can avoid compensating you for something you've got coming, that's more money in their pocket."

Heeding Wheeland's advice, I began carrying a copy of the regulations on every trip. Although they did add to my already heavy flight bag, they came in handy on several occasions when I got into a dispute with an indifferent ticket agent whose compliance with the regulations was inconsistent with the Civil Aeronautics Board's interpretation. As usual, Wheeland's argument was flawless, but then I came to expect nothing less from him.

Wheeland was one of the most professional pilots flying for Globe Aero, an accomplishment in itself, because there were several who prided themselves on their aeronautical skill and knowledge. Wheeland took it a step further. He made it his business to know the regulations that covered every facet of airlines and airplanes. Most of us tried to get by with as little studying as possible and if asked something technical about an airplane we flew often, we would try to change the subject or suddenly remember some urgent errand. Wheeland, however, no matter how big or small the airplane, rattled off performance figures and specifications with such ease that it put the rest of us to shame.

Wheeland's goal was to fly for one of the major airlines, an occupation for which he was eminently suited. Disciplined, confident, handsome, smart as a whip, and, most important, he looked like an airline pilot. In 1975, however, airline jobs were few, and the number of pilots applying for them were many. To prepare himself for this career, at his own expense, he attended a flight-engineer course on the Boeing 727, learning the aircraft and its systems so thoroughly that he knew it as well as the small airplanes he ferried for Globe Aero.

Anyone who can tell the difference between a Boeing 727 and a small Piper knows that about the only thing these two aircraft have in common is that they both operate in the air. Wheeland, however, familiar with both types of craft, saw more similarities, and often pointed them out. When one of us happened to comment on the flight characteristics of the small Pipers and Cessnas we flew, Wheeland would slip into the conversation some piece of information about the Boeing. "It works this way on the 727" or "That's a little like the way the 727 flies" he might say. At the time, the Boeing 727 was one of the few airliners that had three engines and because of Wheeland's frequent references to the airplane, he became known as "Threeholer."

Although we joked about it, it was that same disciplined professionalism and knowledge of arcane facts that saved his and another pilot's life when they flew from Honolulu to Pago Pago and couldn't find the island. Standard procedure at this point was to fly in the same direction for another hour in case they were late because of headwinds, and if still the island didn't show up, turn northwest toward Western Samoa. The first part of this procedure they did exactly as they were supposed to, but then instead of northwest, Wheeland headed northeast toward a thousand miles of open water. They flew until they were down to one hour of fuel, and still nothing. They had their life vests on and were preparing for the almost-certain ditching, when they began receiving the Pago Pago radio station and landed minutes before running out of fuel.

When Wheeland returned to Lock Haven and told us what happened, we naturally questioned his reasons for turning northeast when all logic suggested turning northwest.

"It was just a hunch," he said, modestly.

What Wheeland considered a hunch was not something based on gut feeling, as with most people, but rather a complex set of variables that included ADF nulls, sun position, HF squelch, and a knowledge of wave direction and cloud formations that ancient Polynesian voyagers could've learned a thing or two from. As was his habit, Wheeland enlightened us with a lengthy discourse of all the minute details.

Simplify—that was my motto. Make this job as easy as possible. In keeping with that belief, I perfected my own system of navigation, which I had discovered completely by chance. I was midway between Honolulu and Pago Pago, flying under a clear night sky, and, as usual, with no idea if I was on course, ahead of schedule, or hours behind, when I saw the strobe lights of an aircraft high above me going in the opposite direction. This was not a heavily traveled part of the Pacific, and so seeing what I suspected was an airliner aroused my curiosity and prompted me to call the aircraft on the radio.

The captain identified himself as the Air New Zealand DC-10 on its nightly run from Auckland to Honolulu. We chatted for several minutes before he gave me his longitude and latitude coordinates read directly from his navigation computer. By knowing the airliner's position, I now knew my own position.

For ferry pilots to know their exact coordinates was an event to navigation on a par with the invention of the sextant, or possibly even the compass itself. From then on I began timing my departure from Honolulu so that halfway to Pago Pago I would be able to see the Air New Zealand DC-10. When I figured I was close to the point where we should be passing each other in opposite directions, I would call on the radio until the New Zealand crew answered. Then I would start looking for the strobe lights, and as soon as I saw the aircraft I would ask the pilot for his coordinates. Although this probably didn't do much in the way of making me sound very professional, it did keep down the high cost of navigation.

Professionalism among the small group of pilots that flew for Globe Aero was acquired by two things: the number of crossings one had made and how long they had been ferrying. The gauge for measuring this latter distinction was the flood of 1972. Those pilots hired before and just after the flood—of which Wheeland and Mantzoros numbered in that group—were the senior pilots. While this position didn't pay more, it did mean that they had some choice of what aircraft they wanted to fly or what country they wanted to fly to. For the new pilots like myself, it was more like a lottery. We came into the hangar after returning from our last trip, and if several pilots happened to be in town at the same time, we waited in the breakroom until Moody called us up to his office one by one to hand out the latest assignments.

"I'm going to Sweden" or "Where are you going?" we commented, as each pilot came downstairs after his meeting with Moody wearing either a smile or a frown. As it turned out, the size of one's smile was in direct proportion to the size of the airplane they were assigned, the biggest smile being displayed by those pilots lucky or senior enough to be taking a Navajo.

The Navajo wasn't just any ordinary twin-engine airplane. It was one of the largest airplanes Globe Aero delivered. It was big enough to carry ten passengers in relative comfort close to a thousand miles without extra fuel tanks and was usually equipped with a full set of avionics. It cruised at one hundred seventy knots and could easily climb to seventeen thousand feet, an altitude high enough to be on top of most of the weather and sail along in the lower fringes of the jet stream.

Globe Aero didn't get that many Navajos, and those few we did usually ended up with Waldman in the cockpit. A source of consternation among many of the pilots, they often complained, "I don't care if Phil takes nothing but Navajos, I just don't want him to take all of them."

I should have suspected something when I went into Moody's office, and he asked me if I wanted to ferry a Navajo.

"Sure," I said, excited over the prospect of flying so large an aircraft when my lowly seniority seemed to preclude such an event from happening for several more years at least. I envisioned an easy trip, soaring across the ocean high above the clouds to Kassel, Geneva, or Toussus, when Moody added, "It's only going to San Antonio, but that'll give you time to get familiar with the airplane."

Moody was always working this type of approach with his pilots: offering them a good airplane and then, after they agreed to take the trip, telling them, almost as an afterthought, the bad news that had made everyone else turn it down. So now it was my turn. After flying the oceans to Singapore, Australia, and Europe, I couldn't get excited about Texas. The only consolation was that I would get to see how the airplane flew before taking one over the ocean.

I flew the Navajo by hand for the first hour. After getting a feel for the controls, I sat back in the seat while the autopilot kept the airplane steady at thirteen thousand feet and pointed toward Texas. The last range of the Allegheny Mountains had just passed beneath me when I noticed the illuminated DOOR WARNING light on the annunciator panel.

The Navajo had only one door, which I had closed myself. I was sure it was locked. Because the light came on after I had been flying for two hours, I suspected the problem was a short in the wiring. A warning light, however, no matter how minor, still had to be investigated.

I got up from the pilot's seat and walked to the rear of the cabin. From what I could see, the door appeared to be locked, but I was still bothered by the red warning light. It was possible that in my haste to get started, I didn't slam the door hard enough and one of the six locking pins that kept it closed tight had worked itself loose. That still left five pins; as few as four would probably keep the door closed. But that was only a guess, because I didn't really know what the problem was. Even if it was a pin that had worked loose, if it happened to one, it could happen to the others. And without enough pins to hold it in place, the heavy door could possibly spring open in flight. At the speed I was flying I had no idea how much damage that might do to this brand new airplane.

Maybe I could crack the door open a few inches and slam it closed. It seemed like a simple plan in theory, but first I had to reduce speed. I went back to the cockpit and reduced power before returning to the door. The door was a clamshell type with an upper half, a lower half with three steps, and a flexible cable railing. I turned the handle on the lower door—which locked both halves—with one hand and held on to the upper door with my other hand. Even at the slower airspeed, I felt the slipstream pushing against the weight of the door as I slowly let it fall open a few inches. While occupied with the lower door, I must have relaxed my hold on the upper door, because it began to swing open. As I reached out to grab it, gravity overpowered my hold on the heavy bottom door, which now swung open with a loud whack. Luckily, the cable caught before it could tear itself off the hinges. The cabin suddenly became very noisy.

I had started out with what looked like a minor problem, but now I had a situation that could result in serious damage. I stood there, looking at the mess I had created, and tried to figure out what to do next. I could grab onto the upper door with my left hand and the bottom door with my right hand, but I still needed a third hand to hold onto the airplane. Somehow, I would have to improvise, because I had to get that door closed. There was no way I would attempt to land with it hanging open—the bottom half hung down too low. It would almost certainly scrape the runway when I flared out for the landing, no matter how soft. Since I never landed a Navajo, or any airplane of this size, I didn't have much confidence in my ability to land one so that the word *soft* would be an apt description.

Holding onto the doorway with my right hand, I reached out and grabbed the upper door with my left hand, pulling it down as I wrapped my left ankle around the railing on the bottom door. The top door closed easily enough, but I couldn't lock it without the bottom door, which I tried to pull up with my ankle. The door was too heavy; I had to get a grip lower down on the railing. I slithered down along the edge of the doorway, my body stooped, and tried to maintain contact with the airframe as I lowered myself slowly—mere inches at a time—outside the airplane. I slid my ankle as far down as it would go to where the railing was attached to the bottom step and began pulling up. I held onto the top door with my left hand, my right hand clasped to the doorway, my left ankle tangled around the railing, and my right knee pressed against the inside cabin wall as tightly as I could.

I was half outside the airplane, trying not to look down at the ground two and a half miles below me, when I began pulling the door up with my leg. I didn't have much leverage, but the door was slowly coming up. It was just about where I could grab it with my right hand and close it over the top door. I leaned a few more inches outside the doorway, stretching my arm to where I almost had it, when the airplane began shaking, pitching up, and then slamming down as if someone had gone berserk at the controls.

A stall! I had reduced too much power. With the door open and me acting like a spoiler and breaking up the already reduced airflow, the airplane had lost too much airspeed to maintain flight. It shook as it entered the stall, one wing dropped, and the airplane pointed toward the ground, but then the autopilot caught it and tried to right the airplane. As the nose came back up to the horizon, more airspeed was lost. The airplane started to shake again, a wing dropped, and the airplane pointed itself toward the ground until the autopilot caught hold again.

Each time this cycle repeated, going from a negative G load in the dive to a positive G load in the climb, the doors were pulled open and closed—with me going right along with them. I had to stop these crazy gyrations before the airplane entered a flat spin, from which recovery might be possible, if I wasn't flung outside and left to free fall thirteen thousand feet. Without a parachute it wasn't difficult to imagine how that stunt would turn out.

I twisted my ankle free of the cable, let go of both doors, which immediately swung fully open, and pulled myself back inside the airplane. Groping onto the seatbacks while the airplane pitched and rolled, I staggered back to the pilot's seat, flipped off the autopilot, and pushed the control wheel and throttles forward. At full power the airplane eased out of the stall, the nose came up to the horizon, and the Navajo was back to level flight. I reset the autopilot and let the airspeed increase again before reducing power, although not quite as much as I had earlier.

This time I remained in the pilot's seat until I was sure the airplane wouldn't stall again. Then I went back to the cabin to repeat the entire process with the door, which I hoped wasn't warped from slamming open. Fortunately, it wasn't. I managed to get both halves closed, spending as little time as possible dangling outside the aircraft, and was certain they were locked. I went back to the cockpit, relieved to have this ordeal behind me, sat down in the pilot's seat, and stared at the instrument panel. For all the trouble I went through, the warning light was still on.

I wished I hadn't given up smoking six months earlier, because I needed a cigarette right now. The adrenaline was wearing off, and my knees were shaking. I felt humbled for all my needless efforts, which could have ended disastrously. Strangely enough, I was also amused by the situation. Had I been thrown outside the airplane and fallen to my death, the airplane would continue to fly, probably for several hours, until it ran out of fuel and crashed. When investigators found the wreckage with no pilot on board, I wondered if anyone would make the connection with an unexplained and battered body discovered several hundred miles away. I doubted anyone would and couldn't help but laugh as I pictured the absurd headlines: AIRPLANE CRASHES WITH NO PILOT ON BOARD. POLICE BAFFLED!

I didn't fly another Navajo until several months later. Although the problem with the warning light still cropped up from time to time, I had learned my lesson and would never again investigate it in flight. Besides, the Navajos had more urgent faults in their design that made an illuminated door warning light seem minor in comparison. Faults like a poorly designed heater system that gave me many an uncomfortable crossing and caused me to avoid taking one of these airplanes, especially in winter, in spite of their superior performance.

Ironically, winter, the worst possible time to be flying over the ocean, was when Globe Aero was at its busiest. Our flying, already more than we could keep up with, increased, along with a similar fattening of our bank accounts. No matter how profitable, however, I was also saddened because our numbers were shrinking as pilots left to find work elsewhere.

Bennett was the first to leave, taking a job with Piper Aircraft at their Lakeland, Florida, plant. In the following months, several of the original pilots who had come to work for Globe Aero eventually moved on and were now scattered throughout the country. Campbell and Flournoy left to fly for one of Globe Aero's competitors. Weir, too, was gone, eventually turning up in Pago Pago, once while ferrying an airplane for a competitor and another time flying passengers around American Samoa for South Pacific Island Airways.

Then, in the early months of 1975, the one departure we had all expected became a reality. One day he was in his office sending me off on a trip and, four days later, after I had delivered the airplane and come back to the hangar to turn in my gear, he was gone. After a career that spanned more than thirty years, after flying everything from single-engine trainers to four-engine bombers, Moody retired. In the space of a few days, he had packed all the mementos acquired over the years and moved to Vero Beach, Florida. And now, Waldman was sitting behind his desk and running the whole show.

Waldman's ascendancy to the helm of Globe Aero didn't come as a complete surprise. Every one of us knew Moody planned to retire someday and would most likely sell the business to Waldman. What we weren't prepared for was how fast it happened. What really concerned us was that we didn't know what kind of leader Waldman would make or how this change would affect our jobs. As cranky as Moody was, he was a known quantity we were used to. Waldman, on the other hand, when it came to operating a business, was an unknown. Only the few senior pilots who knew him in the early days when he was just a regular ferry pilot had an idea of what to expect, and none of them were looking forward to working for him.

"Remember, this is the guy who took Donna to the leper colony on Tarawa for their honeymoon," Mantzoros told me, chuckling as he remembered some of Waldman's enterprises. "The guy's a wheeler-dealer. You should've seen him before the flood. He always had some kind of scam going: waterbeds, opals, you name it. If someone wanted something, Phil could get it for them. Customers used to love to see him coming because he'd have all this junk in the airplane."

"I remember him telling me something about traveling around Australia, trading opals," I said.

"That's our Phil. I'd been with Globe for about three months before I even met him, because he was always hanging out overseas after he delivered the airplane. He used to go on a trip and be gone for two or three weeks. Then he'd come back and complain to Walt that he was always broke and he needed a raise. So Walt told him he was spending too much time overseas. 'Do me a favor and try something different,' Walt told him. 'On your next trip, after you deliver the airplane, stay over just one night and come straight home. Then see if you don't come back with more money.' So Phil ran the trip like Walt told him, and sure enough, he came back with a lot more money and started running all his trips like that."

During this time of uncertainty when we were all apprehensive about our future in ferrying, there was one pilot who was amused by the possibility of Globe Aero's financial ruin under Waldman's management. That pilot was Joe Wolf.

Wolf was one of the original pilots who came to work for Globe Aero. He was also one of the fastest. All you had to do was put him in an airplane, wind him up, and, before you knew it, he was back in Lock Haven ready for another trip. He should have been the perfect ferry pilot, except that he was too independent. He operated as he saw fit, and would often get into arguments with the other pilots. Unfortunately, it was not only his peers that he did not get along with. Sooner or later he got around to everyone, including Moody.

Moody would fire him, and then when it got busy again, he would hire him back. After a while, Wolf finally decided he had had enough, because the last time Moody fired him he went to work for Gene Lock. Lock used to fly with Moody when they both worked for Max Conrad. He had operated a small ferry company in the hangar next to ours, which Wolf had recently bought out.

Hoping to build his business and aggravate Moody and Waldman at the same time, Wolf would snoop around Globe Aero's hangar, noting the airplanes and trying to find out from the pilots where we were going and what new customers we were delivering to. Then he would contact the same customers and quote them a price for delivering their aircraft that, invariably, was less than Globe Aero charged. Today, however, Wolf was on a dual mission. Besides customers and airplanes, the first words out of his mouth were, "So Phil's the big kahuna now, huh?"

Wolf was chunky in both physique and personality, with an expression somewhere between a grin and a sneer that seemed to be permanently frozen on his face. He had known Waldman longer than any of the other pilots, and his speculation made everyone else's seem like high praise in comparison. "I hope he does a better job with airplanes than he did with wigs," he said, grinning wide.

"Oh, Jesus, the wigs," Mantzoros said.

"What wigs?" I asked. "I thought I'd heard all of Phil's scams."

"No one told you about when Phil was selling wigs?" Wolf asked, pleased that his visit was causing such dissension in the ranks.

"It was back when ferrying was slow," Mantzoros said. "Everybody was doing something else between trips: flying charter or flight instructing. And Phil was selling wigs."

"He sold a lot of them, too," Wolf said, chuckling as he remembered this chapter in Waldman's career. "But you have to realize, they weren't exactly top of the line."

Mantzoros and Wolf looked at each other and both started laughing as they recalled the incident that put an end to any hopes Waldman might have had of cornering the Lock Haven wig market.

"He did great, all right," Mantzoros said, "until some lady came to the hangar one day looking for Phil and mad as hell because she was wearing one of his wigs in the rain and the colors ran all over her."

Over the next few months the exodus of pilots continued. First John Snidow left to take a flying job with Air Bahamas; and then Mantzoros followed Bennett to Piper Aircraft. These departures, I surmised, were prompted by the feelings of uncertainty that dominated our thoughts and conversations. Without Moody's firm hand at the helm, we feared Globe Aero would be rudderless, drifting, with no clear definition of where it was going. I tried not to pay too much attention to all the rumors and suspicions, but it was difficult. There was talk of mutiny, mass walkouts, and storming into Waldman's office at the first indication that things were getting worse. The future looked gloomy, yet there was nothing we could do but wait and see.

(*Left to right*) Dave Gray, Andy Mattenheimer, and me somewhere in the South Pacific, March 1977.

(*Left to right*) Jack Seaman, Ernie Kuney, and Ernie Wheeland waiting for winds in San Francisco, 1973. (Courtesy Ernie Wheeland)

Piper Arrow and Beech Bonanza on Norfolk Island en route to Sydney, Australia, 1974. (Courtesy Ernie Wheeland)

Aero Commander on a snowy ramp in Goose Bay, Labrador, 1975. (Courtesy Ernie Wheeland)

(*Left to right*) Donn Kerby, Andy Mattenheimer, Paul Briggs, me, and Dave Gray with one of the Grumman Albatrosses we ferried to Surabaya, Indonesia. The second Albatross, which was piloted by Gray, Mattenheimer, and myself, was left on the island of Morotai after flying 500 miles on one engine, September 1976.

Andy Mattenheimer (*right*) and me in the cockpit of a Grumman HU-16B Albatross preparing to depart for Indonesia, September 1976.

Somewhere over the Pacific in two Grumman Albatrosses that Paul Briggs, Donn Kerby, Dave Gray, Andy Mattenheimer, and I ferried to the Indonesian air force in Surabaya before the aircraft began showing their age, September 1976.

(*Left to right*) Jim Smith, Bob Moriarty, me, and Tim Peltz with one of four Ayres Thrush S-2R crop dusters en route to Barquisimeto, Venezuela, February 1977.

(*Left to right*) Phil Waldman, Bob Campbell, Dave Gray, and me at the General Aviation Manufacturers Association Convention, Orlando, Florida, 1977.

Three Piper Aztecs that Tim Peltz, Bob Moriarty, and I ferried to the Nigerian air force in Kano, February 1977. (Courtesy Globe Aero)

(*Left to right*) Jim Bennett, me, Dave Gray, Donn Kerby, and Nils Mantzoros at Gray's wedding, Ormond Beach, Florida, 1978.

Scraping ice off a Piper Cadet parked under the wing of a Lockheed C-130 Hercules at St. Johns, Newfoundland, December 1988.

Beech Bonanza piloted by Clark "Woody" Woodward somewhere over Africa, September 1989.

Resting on the wing of a Beech Bonanza somewhere in the Far East, early 1990s.

(*Left to right*) A client and me in front of the Cessna Caravan I had just ferried from Paris to Godthab, Greenland, August 1992.

A Piper Malibu parked next to the passenger terminal at Narsarsuaq, Greenland, early1990s.

The fjord at Narsarsuaq, Greenland.

(*Left to right*) Jon Egaas and I enjoy a few days of crew rest on Wakiki, Honolulu, May 1989.

The Beech F-33C Bonanza that I crashed onto a farm when the engine quit just after taking off from Lakeland, Florida, in December 1990. In more than 300 ocean crossings, this was the second airplane that I crashed—oddly enough the first one was also a Bonanza.

Ernie Wheeland during a Piper Navajo trip to Bankstown, Australia, early 1970s. (Courtesy Ernie Wheeland)

Bill De Long and a Piper Arrow he was ferrying to Europe—note the temporary ferry fuel tank where the right front seat used to be—Shannon, Ireland, 1973. (Courtesy Ernie Wheeland)

Piper Arrow with an ominous registratioin number that Andy Knox ferried to Germany in 1978. Both plane and pilot arrived at destination safely. (Courtesy Andy Knox)

9

Into the Revolution

Waldman surprised us all. Instead of running the company into the ground, he turned out to be a pretty good businessman. He was also easier to work for than Moody, especially when it came to handing out trips. And he was almost honest. He underwent such a complete change of personality that I had a hard time believing the person I had known for the past year was the same one sitting in Moody's old office now.

One of the first things he did as the new owner was to increase our pay. Although it still lagged behind other companies, the gap wasn't as wide as it had been under Moody. In fact, we probably earned more because over time we flew more trips. Already larger than any of our competitors, Waldman wanted to make Globe Aero bigger still. He went after new business with a fervor I never would have thought him capable of. Before long we were ferrying airplanes to new customers in Europe and the Pacific. I was flying to someplace new on the average of once a month, and was now on my way to South Africa.

"I've got just the trip for you," Waldman said, when I went into his office to find out where I would be going next. "We've got this new customer,

Comair. They're the Cessna distributor for southern Africa, and they want us to take this one airplane as a kind of test. So if we don't screw up, they'll probably have a lot more for us. That's why I want you to take it, because I know you'll do a good job getting it there quick and in one piece."

Since taking over the company, Waldman had turned into a schmoozer. "You're the only one I can trust to take this airplane," he would say, or, "If you get there by next week the customer says he'll fix you up with his secretary." Usually this turned out to be a hot date fabricated in Waldman's mind only, but he would say anything to get someone to take a trip.

"What airplane?" I asked, having learned from Moody that this information should be ascertained before committing to anything.

"Good old Africa," Waldman said, carefully avoiding what he didn't want to talk about until I was sufficiently hooked. He began by telling me about when he used to fly to Africa in the late 1960s, land at Sao Tome, and go drinking with the mercenaries fighting in the Biafran War. "Now there was a bunch of guys that lived hard, fought hard, and played hard," he said, recalling the good old days when he was a simple bachelor ferry pilot without a care or responsibility in the world. About the only things he left out were white hunters and diamond smugglers. He also failed to say anything that could be of any real use.

"But what can you tell me about flying there?" I asked.

"If you like challenges, Africa's the place," he said, and then briefed me on what little he remembered, beginning with the weather: "It can get pretty rough over the desert, but if you stay over water, about a hundred miles from land, it shouldn't be too bad." Navigation: "Radio beacons are scarce, and a lot of them aren't turned on." Where to stop for fuel: "Abidjan's the best place; it's cheap and pretty stable. Everybody stops there." Politics: "Try to avoid flying over big cities, because they get nervous with unscheduled flights." He finished with something about a civil war in Angola.

Sure I would go, I told him. It sounded too adventurous not to go, even after he finally got around to telling me that I would be taking the Cessna 182 parked outside the hangar.

"You mean that small, single-engine Cessna?" I asked, my enthusiasm waning. "The one that'll probably do a hundred twenty knots."

"You're being too negative. It'll get up to a hundred twenty-five, at least."

Africa

"Oh, excuse me. A hundred twenty-five. It'll still take two weeks to get there."

"But you'll have company. Rice is taking another Cessna to Tanzania. You can fly most of the way together."

I was not so sure that this was a blessing. If there was one thing everybody at Globe Aero was in agreement with, it was that a dark cloud seemed to follow George Rice wherever he went. Almost from the moment he left on a trip the problems would begin. It might be a freak blizzard or typhoon that materialized out of nowhere and prevented him from flying any farther. Or maybe his airplane was falling apart, his compass had gone haywire, he was coming down with scurvy, or a war had just broken out. It got so that whenever Rice left on a trip, it was only a matter of time before Waldman got a phone call from him about some new and dreaded malady that was causing his latest delay.

"They're bombing the hell out of each other," he had told Waldman a few months earlier, when fighting erupted in the African country he was about to fly to. "You have to get me clearance to fly somewhere else. Either that or leave the airplane here 'til they get their act together."

Either Waldman had gotten too many such phone calls or his immediate priorities in Lock Haven were different than Rice's in Africa, because his response was not one that Rice was looking for.

"I don't know what you're worried about. The airplane's insured."

Although Rice got in and out of the country without incident, it was not before one of the airport controllers took him to the morgue because he wanted an American, any American, to see all the killing that was taking place in his country.

Rice's problems were endless, and, whether real or imagined, they were problems that no one else seemed to have or that had ever happened in the history of aviation. But they happened to Rice so often that the rest of us were afraid his bad luck was so contagious it would cause havoc to anyone unlucky enough to be in the same ocean.

"I see Phil's conned you, too," he said, when we hooked up at the Albatross Hotel in Gander. Rice was short and looked young for his age. He had an unruly mop of curly red hair that no matter how many times he combed it, it did exactly what it wanted. The first thing I asked was if he was having any problems with his airplane.

"I don't know. It seemed to be running a little rough at first, but I guess it's okay," he said, not too enthusiastically.

Running okay was about as optimistic an evaluation as I figured I was going to get. With no excuse to delay, we began planning our route.

"We probably ought to try and stay together as long as we can," he said.

"Yeah. That way if one of us gets shot down, at least someone will know about it," I said, only half joking. "Maybe we can fly to Angola together; then you can head to Tanzania from there."

"And fly across the jungles of Zaire and Angola? Not this kid. Oscar Lima thinks this trip is dangerous enough. She'd kill me if she thought I was making it worse."

Oscar Lima, the phonetic alphabet for the letters *O* and *L*, in Rice-speak stood for "old lady," an acronym that failed to do justice in describing his Pan American stewardess wife.

"You know," Rice continued, "I was looking at the map and, oddly enough, the most direct route for me is through Shannon, across Europe, and down the east coast of Africa."

"It's going to be hard for us to run this trip together then, since I should go to the Canary Islands from here. That's about a thousand miles from Shannon."

"Maybe we can find some place in between to go to," Rice suggested.

"Even if we split that in the middle, it's still five hundred miles out of route. Walt would've gone ballistic if we went even fifty miles out of the way."

"But Phil says we should fly together for part of the trip," Rice said, as if this was a mandate with unlimited restrictions. "And I think we ought to do like he says."

"But fly to where? Every airport in southern Europe or North Africa is either too expensive or doesn't have av gas."

"We could fly to Lisbon," he said.

"Lisbon? It's probably expensive, they're not geared up to handle small airplanes, and I can't think of one possible reason why we should fly there. Can you?"

"Yeah. We've never been there before."

So we left for Lisbon on a direct course that took us within a few hundred miles of the Azores, a small chain of islands two-thirds of the way between Newfoundland and Portugal. Santa Maria, the main island in the chain, would've been an ideal place to land and break up the leg if the Portuguese weren't so strict about HF communications. Basically, they wanted the radios to work. They wanted position reports every hour and tolerated nothing less. Given the unreliability of HF communications, most ferry pilots avoided the Azores, especially after Slim Byrd, one of the many independent ferry pilots at the time, decided to test the limits of their tolerance.

Upon landing at Santa Maria, Byrd was promptly informed that unless he got the radio fixed he would not be allowed to leave. Two days later, with the radio still not working, Byrd apparently decided he had hung around long enough. Just after midnight, he went out to his aircraft, started the engine, and, without bothering to call the control tower for clearance, taxied out to the runway and took off. When he landed in Tenerife five hours later, the Spanish controllers knew all about what had happened in Santa Maria. Although they got a good laugh over the trick he played on the Portuguese, they also warned Byrd that he better never go back there again.

Due to the condition of our radios, Rice and I stayed well clear of the Azores and relayed our position reports through the airlines. It never occurred to us that if the Portuguese in the Azores were strict about communications, the aviation authorities in their home country might be just as unyielding.

Fortunately, they weren't. They didn't seem to care if we talked to them or not. After we landed they parked us in the outer perimeter of the airport, half a mile from the terminal, and busied themselves with the Boeings and Airbuses. Eventually a shuttle bus came along and drove us to the operations building, where we began looking for the flight-planning office.

Everywhere we went it was the same: I stopped at a counter to ask directions, but people kept right on working as if I didn't exist. Then Rice asked, and they fell all over themselves to give directions. When we came to a closed door, security guards rushed to open it for Rice and then stopped me to ask for identification and wanted to know what I was doing there. I swear the last guard even saluted him. Sensing my frustration, Rice finally suggested that any preferential treatment he was receiving was because of his shirt.

"What *about* the shirt?" I asked, looking at Rice's white pilot shirt with epaulets and flaps over the pockets and then at my plain blue shirt.

"You have to psyche these people out. We're not that far from Africa. If you want to get around airports in this part of the world without any hassles, it's not enough to be a pilot, you have to look like a pilot."

Rice was of the opinion that what people believed to be true was more important than truth itself. Often, he registered in hotels using the name Colonel Rice because, he claimed, it kept people wondering who he really was. With his disarming manner, he seemed to fit in wherever he went as if he had been coming there his entire life. He spoke no language other than English, yet somehow conversed so that both he and the person he was talking to knew what was said. Despite any dark clouds or unknown disasters his presence might cause, I was sorry when we split up the following morning. He went to Crete and the east coast of Africa, while I headed south, toward Africa's west coast. As Waldman had suggested, I flew a hundred miles off the coastline and made landfall over Mauritania, just south of the Spanish Sahara.

In this part of the world, where haze reduces visibility to one or two miles, my arrival over the continent was decidedly anticlimatic. One minute I was flying over water, and the next minute I was over land—light brown, flat, and boring—that went unchanged for hours. Occasionally I spied a small coastal village through the haze and then, after the sun dipped below the horizon, dim clusters of lights that made the darkness seem even more forbidding.

Even with the lights, I didn't see Dakar until I was almost over the city and then had to fly past the airport a few miles so I could descend without dive-bombing toward the ground. From then on it was chaos. The African controller spoke English, but with a French accent that sounded like one long melodious word of alternating peaks and valleys. In spite of this, I managed to get the airplane on the ground without incurring the controller's wrath. Given the number of airliners taking off and landing, I considered this a major accomplishment.

Dakar rests on the extreme western tip of Africa, where the continent juts out and points to South America. Eons ago it might have been the eastern end of the land bridge that connected Africa with the Americas before the continents drifted apart and the Atlantic Ocean was formed. Now it was part of the air bridge used as a way point for European airlines bound for South America and the occasional ferry pilot who, like me, tried to stay as inconspicuous as possible.

Waldman had warned me about flying in black Africa in an airplane that was ultimately going to white Africa. "Just don't draw attention to yourself," he had said. I didn't, until I went to pay the landing fee and was presented with a bill that had several numbers with a string of zeroes behind them. "I hope this isn't dollars," I said.

"No, captain. It is Central African francs," the African clerk said. "You pay in dollars?"

"Yes," I said, going into my Pigeon English. "I pay you in dollars if you convert."

He converted the amount into a smaller and more manageable figure, but one still much higher than I had paid anywhere else. "This is a lot of money for such a small airplane," I said. "I don't want to *buy* the airport."

"But, Captain, you must pay for runway lights."

"Runway lights? What're you charging me for runway lights for?"

"Because you land after it is dark."

"But the lights were already on."

"Yes. So you pay."

In the end, I paid. It was late, and, if I wanted to get any sleep, I figured I had better find a hotel soon. With the clerk's help, I called all the big hotels near the airport—which were all booked for the night—before ending up in downtown Dakar at a small local hotel, where the desk clerk warned me several times to double lock the door. My first night in Africa and, other than the expensive landing fee, there had been no major surprises. Today had been the easy part. I still had four thousand miles to fly, unless I figured out some way to shorten it.

The following morning, after waiting until the runway lights were turned off, instead of flying a hundred miles off the coast, I stayed over land and flew direct to Abidjan. Although the chart showed an airway between these two cities, there were only two navigation beacons, one at Dakar and the other at Abidjan. There were several VOR stations a hundred miles or so on either side of the airway, but none that I would be in range of if I stayed on course. So I used them in reverse, keeping my radio tuned to one station, and if the needle started moving, indicating I was getting close to the station and off course, I turned a few degrees in the opposite direction. Like tacking a sailboat, I zigzagged for the next several hours, being careful not to venture into the wrong airspace, until I began receiving the Abidjan VOR.

I picked my way around small cumulus clouds as I descended through the haze over the jungle. The temperature outside the airplane was twenty degrees Centigrade, and I still had five thousand feet to descend. As soon as I landed and turned off the runway I opened the side window, but at three hundred miles north of the equator, all that did was mix the stale, hot, and sticky air inside the airplane with hot and sticky air from outside the airplane.

Waldman was right about Abidjan. Everybody did seem friendly, and it was a busy place. No sooner did I finish refueling, than two other small airplanes landed and taxied over to where I was parked. The HF antennas dangling from the belly of their aircraft gave them away as also being ferry flights.

"Welcome to Africa," said one of the pilots, as he walked over to me and stuck out his hand. "Gerry Swoboda. Where're you going?"

This, more than anything, was the universal greeting among ferry pilots. First the welcome to wherever you happened to be, then the question of where you were going, and, eventually, what company you flew for. It was like a secret handshake.

"South Africa. How about you?" I replied.

"Gabon," said Steve Hall, the pilot of the second airplane, as he walked over and introduced himself. Swoboda, meanwhile, had taken five passports all bearing his picture, out of his suitcase and was trying to decide which one he should use.

"Why don't you be a Spaniard?" suggested Hall.

"I was Spanish yesterday in Villa Cisneros. I think I'll be Argentinean today. I haven't used that passport in a while."

With this problem solved, Hall turned to me and asked, "Where're you staying?"

"I don't know yet. I've never been here before."

"This your first time in Africa?" Swoboda asked.

"Yeah. What can you tell me about it?"

"Well, if you come here often enough, sooner or later, you're going to end up where you shouldn't be. What you want to do is fly at night with your position lights off, don't talk to anyone, file a phony flight plan, and give false position reports. That way, if they send fighters after you, they'll be looking in the wrong place."

"Won't they get worried when you don't show up at the place you filed to?"

"They might, if Abidjan sent them the flight plan. But I really don't think anybody talks to anybody down here," said Hall. "Most of these countries are anti-American, anti–South African, or just plain antieverything that comes out of the West. Abidjan's probably okay, as long as you just get in and out and don't let too many people know you're here. So you don't want to go downtown. In fact, don't go through customs unless you want a lot of hassles. They've got a few hotel rooms upstairs in the terminal. Why don't you get a room there? We'll be over as soon as we finish refueling and meet you for dinner."

Hall and Swoboda worked for one of Globe Aero's competitors, Brand X, we called it, the same name I'm sure they used when referring to us. When I joined them in the restaurant, Hall was recalling the time he was exchanging Canadian dollars for Central African francs at the bank in Abidjan's airport terminal.

"The guy kept giving me all this play money," Hall boasted in a deep beer-barrel voice, a voice that could cheer the home team on to victory or urge everyone to leap, like so many lemmings, into a deep chasm while he stayed safely on the precipice.

"Way too much money," he said, grinning. "A lot more than I figured the Canadian dollar was worth. So I asked him if he was sure about the exchange rate. 'Yeah, Yeah,' he says, trying to sound real important. 'This is the official rate today for English pounds.' And he kept shoving more money at me. Then it dawned on me. He must've seen the Queen's picture on the bills and thought they were English pounds."

"Whoa. Big difference," said Swoboda. "What'd he say when you told him about it?"

"What're you nuts? I didn't say anything. I started going through my pockets, looking for all the Canadian money I could find."

"So they ripped you off in Dakar," Swoboda said, when I told them about the runway-light fee.

"All the ex-French colonies charge for runway lights," said Hall. "Doesn't matter if it's light out or not. As long as the lights are on, they're going to charge you. If you have to pay, exchange your money first and pay in African francs or else they'll give you a lousy exchange rate."

"Now in Angola," said Swoboda, "except for the hotel, the last thing you want to do is pay in escudos."

"Why's that?" I asked.

"Because everyone's leaving the country, and they want hard currency."

"What's this about a civil war, anyway?"

"Independence," Hall said. "The Portuguese are pulling out of Africa. The only problem is, they didn't turn the country over to anyone. So now you've got these three factions all fighting for power. Luanda's okay right now, except the government and the whole damn infrastructure is falling apart."

"The airport's even worse," said Swoboda. "The place is crowded with everyone running around, and here you are with all these things you have to do: customs, fuel, flight plans . . . you know . . . all the normal stuff."

"Anyone getting shot at?" I asked, with some concern.

"Ah, I wouldn't worry about that," said Hall. "The real problem is trying to get a taxi to the hotel. What you have to do is hire yourself a pimp. They're all over the airport. Someone'll just come up to you and offer to be your agent."

"I suppose this'll cost a fortune."

"A couple of dollars is all," Swoboda said. "Everything's a couple of dollars."

Armed with what I hoped was current information about the war, and not just ramblings spurred on by drink and macho camaraderie, I left the following morning for Angola. Hall and Swoboda departed for Libreville in Gabon. The briefing I had gotten from Waldman, while better than no briefing at all, was only slightly better. A lot had changed in the half dozen or so years since the Biafran War. I had come this far without worrying about what those changes might be, but now that I was getting closer to Angola, the reality of flying into a revolution was becoming a matter of concern.

It was just after sunset as I approached the coastline, much clearer without the haze that permeated northern Africa. I saw the lights of Luanda long before I began the descent. I was thankful, however, less for what I saw than for what I didn't see: tracers or bombs going off. I was grateful for this because I didn't have enough fuel to go anywhere else.

On the ground the airport looked like any of the other African airports I had been to. A TAP Boeing was parked on the ramp, along with several other airliners whose names I didn't recognize. Fuel trucks and Follow Me jeeps wove their way between aircraft and mountains of cargo containers stacked near the terminal. I refueled quickly, gave the soldier standing guard a few dollars to watch my airplane overnight, and hoped he wasn't one of the rebels.

I paid for everything in U.S. dollars and then bought escudos from the Portuguese control tower operators, who, as Hall had said, gave the best black-market rate.

"There will be nothing here for us after Independence," said the controller. "We will go back to Portugal, and outside of Angola the escudo is worth nothing."

Everybody I ran into wanted to sell escudos. It was a buyer's market, a crazy bazaar, where the Portuguese residents exchanged their soon-to-be worthless script for paltry amounts of hard currency. Currency they hoarded until the day they would be kicked out of the country by the new government.

Who that government would be was anybody's guess. The three nationalist groups—the Popular Movement for the Liberation of Angola (MPLA), the National Front for the Liberation of Angola (FNLA), and the National Union for the Total Independence of Angola (UNITA)—were fighting a civil war for control of the country. They hid in the jungle and mounted guerrilla raids against the small villages. It was getting worse every day, fighting was escalating and moving into the cities. Independence Day was getting closer, and the capital city of Luanda was being hit with random bombings. The acronyms MPLA, UNITA, and FNLA were spray painted on the bullet-riddled walls of the nearly deserted terminal, along with the words *Indepencia* and posters. The poster that caught my attention was of a clench-fisted guerrilla leader named Jonas Savimba, which I bought from the security guard for a few dollars and a pack of cigarettes.

Angola was what I imagine Indochina was like just before the fall of Saigon, a powder keg ready to burst apart at the seams. Outside on the street side of the passenger terminal Africans milled around, talking loudly in Portuguese and what sounded like African dialects. There were a few Europeans, seemingly lost in the sea of Africans whose angry faces looked as if they might be one step away from mob hysteria. I felt a sense of urgency, as if the people, the buildings, and the cars and buses belching exhaust smoke were about to explode. I made several attempts to flag down a cab, but every one of them drove past as if I was invisible.

"You wanna taxi?" said an African who seemed to materialize from out of nowhere. He was wearing a shiny sports jacket, pointy shoes, and a porkpie hat—obviously the pimp Hall and Swoboda said would come to my rescue.

"Yeah," I said, and before we settled on a price, a taxi was braking to a stop in front of us. A minute later we were speeding into the city, down wide avenues lined with bombed-out buildings, past street merchants with wares set up on dilapidated carts and farmers walking their scrawny livestock alongside dusty streets.

The Le Presidente Meridien Hotel looked as if at one time it might have catered to affluent European businessmen and tourists. Any claim to grandeur, however, was entirely in the past. The palm trees lining the wide drive and the once-manicured lawn were growing wild with neglect, four stars were prominently displayed on a cracked and peeling stucco wall, and the intricately patterned tile floor of the lobby was chipped and worn. Desk clerks and bellmen still attended to guests promptly, but now their tuxedos were soiled and frayed.

I had a late dinner in the hotel restaurant, a place frequented by hookers, businessmen, and what I suspected were intelligence agents trying to pass themselves off as businessmen. The dogs of war, as Shakespeare warned, were all in attendance, plotting to capitalize on the misery of this unfortunate country.

As Hall and Swoboda had said, all too often there was a civil war or coup d'état somewhere in Africa. International borders changed, governments changed, and political allegiances changed so fast one could barely keep up with them. As a result, the political face of the continent was constantly being redrawn. Many countries had only recently shed the bondage of colonialism and were now going through the political adolescence of military juntas, dictatorships, and one-party rule. The cold war between the superpowers was at its peak, and Africa, like Southeast Asia, was the arena where the success of their rhetoric could be measured in geographic terms. Countries were either in the Eastern or the Western camp. Puppet rulers got away with almost any atrocity as long as they remained loyal to the ideologies of their superpower benefactor.

Africa, first and foremost, is a land of paradox. While it is understandable that a continent and a people going through the upheaval of political change would not always be what they appeared to be, I still found the difference between black and white Africa ironic. Angola, the country I thought would be difficult, was easy. But South Africa, the country I thought would be easy, turned out to be difficult. It was no one's fault but mine, because I overslept in Luanda. It was mid-afternoon by the time I reached the Kalahari Desert and then had to dodge thunderstorms for the next several hours. Because of this, I didn't get to Johannesburg until after dark and then ran straight into the biggest thunderstorm I had ever encountered.

I had seen cumulonimbus clouds in the Pacific, but none so huge as this that it soared to thirty thousand feet and covered my entire field of vision. It stood defiantly, like a medieval fortress with great parapets of clouds that looked so forbidding as to repel the most intrepid invader. At ever-increasing intervals, lightning split the heavens, turning night into a day so crystal sharp I could clearly see roads and buildings of the city underneath.

Because Rand Airport, the small, general aviation field where I was supposed to deliver the airplane, was closed, either because of the storm or the late hour, approach control directed me to land at Jan Smuts, the international airport on the outskirts of the city.

I had just parked the airplane when a white Afrikaner customs officer walked over, eyeing me like a traffic cop, and waited for me to climb out before barking, "Do you have your passport?"

I handed over my passport and watched while he studied every page, thumbing over the bureaucratic stamps as intently as if he were searching for some clue that would blow my cover of being a spy or a terrorist. I was beginning to suspect something was wrong, that maybe I was someplace I shouldn't be. I had had this feeling ever since I landed in Africa, but figured paranoia was normal when flying in this part of the world.

"You have the keys to your airplane?" he snapped. It may have been phrased like a question, but I knew it wasn't. He *wanted* the keys.

"Sure," I said. It was raining, so I figured he was angry because he had to come outside and decided to take it out on me.

He locked the airplane, put the keys in his pocket and snapped another command. "Come with me."

Because of the rain, I figured he just wanted to clear me into the country tonight, and we would sort out the airplane papers tomorrow. If this was the case, he didn't tell me. He didn't say anything. He just started walking toward the terminal, leaving me to wonder what was going on. The officer had let me take one suitcase, the one with all my clothes, but everything else—charts, radios, and emergency gear—were still in the airplane. Something didn't seem right. "You mind if I get my gear out of the airplane?" I asked.

"No. The airplane is locked up," he said, sternly.

Now I was beginning to get angry. "Hey, what's the problem?" I asked.

"The problem," he said, looking at me as if I was a criminal, "is that you do not have a visa for South Africa."

Big deal, I thought. South Africa was part of the United Kingdom, or was at one time. It was just like Australia, and I didn't need a visa there as long as I went in as an aircrew member. "I'm aircrew. I don't need a visa," I said, with confidence.

"This is South Africa," he said, apparently unimpressed with my answer. "Everybody needs a visa or you cannot enter the country."

"Well, I'm here now. So what are we going to do?"

"What we are going to do is put you on the next airline that is leaving the country," he said, and kept on walking.

"No, really," I said, thinking he was kidding, although he didn't look it. "What's going to happen?"

He repeated my punishment.

"What? I just got here! I just flew all the way from the United States. Can't you give me a temporary visa or something?"

"To do that, you have to be in South Africa."

"Well? That's where I am," I said, figuring I had him now. "So stamp my passport."

"I cannot do that."

"I don't understand. Why not?"

"Because without a visa, you cannot enter the country."

"Let me see if I've got this straight. You can't give me a visa because I don't have a visa to enter the country. But if I did have a visa, then you could give me one."

"Exactly."

The guy was serious. And now he was going to kick me out of the country because of some inane technicality. Any more arguing, I could see, was just a waste of time. All I wanted was to get my things out of the airplane, and they could send me wherever they wanted. "If you're going to kick me out of the country, at least let me get my gear out of the airplane."

"It is too late to do that now. The airplane is sealed."

"Can't you unseal it? All my radios and charts are in the airplane."

He spoke with another customs agent in Afrikaans and then back to me, repeating the same official line about the aircraft being sealed and that I should have gotten a visa before coming here. I was getting nowhere. The more I argued, the more frustrated I got. He did let me phone my customer so I could let them know the airplane was in South Africa and that I, as he put it, was in a lot of trouble.

One hour later, a white South African came into the customs office and appraised me up and down as if trying to decide if I was the guilty party. Because I was the only one in the room wearing civilian clothes it was an easy guess.

"Boy, am I glad to see you," I said, after he walked over to me and introduced himself as Jimmy Morant. An English name, I thought. Finally, I had some help, someone on my side.

"Seems like you've gotten yourself into a bit of trouble here," he said. "Why don't you tell me what happened."

As I told him everything that took place since I landed, I found myself getting angrier and not doing a very good job of disguising it. "And now these Nazis want to kick me out of the country."

"Okay, you better let me take care of this. Just wait here and don't say anything."

I waited while he talked with the customs agents. There were several involved by this time, all speaking in Afrikaans. Twenty minutes later, they gave me back my passport along with a sheet of paper stapled to it, authorizing me to "visit the Republic of South Africa for a period not to exceed twenty-four hours." It also stated that Comair was responsible for my actions while in the country. I made a mental note to not be so hasty with pretrip planning in the future.

What struck me as odd the following morning when I went for a short walk outside of the hotel, was that Johannesburg didn't seem to belong on the African continent. The Africa I had seen so far was small villages in the jungle or flat cities in the desert, thatch-roofed huts, dusty streets, and a people caught in an unending cycle of poverty. In contrast, Johannesburg was a modern, cosmopolitan city with gleaming skyscrapers, luxury hotels, and trendy neighborhoods with art galleries and outdoor cafes. For a big city, it looked almost too clean.

"That is because the Dutch Reform Church keeps a tight hand on everything," Morant told me, when he picked me up at the hotel and took me to the airport to collect the rest of my gear. "There is no television in South Africa. No gambling or pornography of any kind. You cannot even buy a *Playboy* magazine."

"Now that you mention it, I don't remember seeing a television in the hotel room. So what do you do for entertainment? You must have a lot of nightclubs."

"We call them private clubs. You see, everything is discreet here. If people want more, then they go to Sun City, the big gambling resort in Bophuthatswana. It is like your Las Vegas: big casinos, nightclubs, strippers, you name it. On Friday night, everybody goes there to get drunk, gamble, and do all the things they cannot do in South Africa. Then Monday morning they come back to their jobs like good citizens, until they can go back next weekend."

"Isn't Bophuthatswana part of South Africa? According to the map, it's right in the middle of the country."

"Yes, but it is one of the homelands. That is the National Party's plan to deal with the black problem. After South Africa became a republic in 1961, they created these homelands so they could relocate all the blacks there. They are citizens of the homeland; they live there, but travel to South Africa every day to go to work. And because they are independent countries with their own governments, they need a passport to travel back and forth."

"And this plan is working?" I asked, skeptically.

"Not very well. The only country that recognizes them is South Africa. But the National Party and all those Dutchmen who run the country plan to create more—nine, I believe."

"I take it you're not too impressed with the Dutch Afrikaners."

"I am English. We fought a war with them in the beginning of the century, and we've been fighting with them ever since, only now we do it in Parliament."

After taking me back to the hotel and reminding me not to miss my airline flight that evening, Morant left me to explore the city by myself. Despite everything I had just learned, I tried to remain objective, but it was difficult. Everywhere I went there was a policeman. Unlike London's bobbies or Paris's gendarmes, these cops had enough riot gear—submachine guns slung over their shoulder and German Shepherds that growled whenever a black man walked by—to make the toughest urban terrorist or petty criminal think twice. I couldn't go around a corner without seeing another one. Even in the Carleton Shopping Centre, a vast underground labyrinth of fancy shops and restaurants that claimed to be the largest indoor shopping mall in the Southern Hemisphere. It was there, amidst the displays of diamonds and gold, where one could buy anything from elephant-hide purses to the latest European fashions, that I saw the signs directing shoppers to the restrooms: one for Europeans and one for Africans. It was obvious that the distinction was based on color, not nationality.

The Republic of South Africa was a country whose climate and geography, not unlike the United States or western Europe, produced an abundance of lush forests, crystal clear lakes, sandy white beaches, vast deserts, and majestic mountains. It was one of the most stable and democratic countries on the continent, yet beneath it all was the social and racial baggage of a small southern town in America before the Civil Rights Movement.

If I had any questions about what all of the whites—both Anglo and Boer—thought about this policy they called *apartheid*, they were answered later that day when I met a particular white Afrikaner in the Carleton Hotel's bar. Alex Mathius was a mining-equipment salesman from Durban, who figured he had had just about enough of all the racial problems in *his* country.

"They are not like the blacks in your country," he complained. "Over here, they are all ungrateful. We give them jobs and a place to live, but is that enough? No. At the end of the day, they want more."

"What do they want?" I asked. I tried to settle the heavy feeling in my stomach from the oxen sausage and boiled tongue I had for lunch with more Cape wine.

"Now they want an equal say in the government. Well, we will not let that happen."

"There's twenty million blacks and four million whites in South Africa; sooner or later there's going to be majority rule."

"But they are not a majority. They are all from different tribes. All this killing you read about, that is them killing each other. And everybody blames the government. These Kaffirs you read about that are arrested, and then go crazy in their cell and start banging their head against a wall until they kill themselves—and the police try to stop them, but what can they do? So everybody yells 'police brutality.' I say let them go ahead and kill each other."

No matter what I said, I was not going to change his thinking. There was too long a history of hate between South Africa's white and black races. Mathius may have been looking right at me, but in his mind he was probably seeing his Boer ancestors who left their homes to embark on the Great Trek. It was the same ancestors who prayed for victory when Voortrekker leader Andries Pretorius and a commando of 464 men fought thirteen thousand Zulus on the banks of the Ncome River. By the end of the day, the Zulus had fled, leaving three thousand dead and the river dark with blood. Not one Boer died, giving rise to the Afrikaner myth that their Old Testament God had protected them against invincible odds to prove to them that they were His chosen race. That day in 1838 is still held sacred today in memory of the Battle of Blood River.

While this Boer's feelings about the racial problem were extreme, it was understandable—he felt threatened. Because of cheap black labor, most South Afrikans enjoyed an excellent standard of living, which made it difficult to change. As for the blacks not being a majority, I didn't want to agree with him, but I wasn't entirely convinced that he was wrong. Like much of the continent, South Africa's black population was made up of different tribes and political groups that hated each other as much as they hated the white Afrikaner.

I was curious about this country that seemed to have so much history in common with the United States, a country whose native population had been driven off their land and segregated just as the Zulus, the Bantus, and the Xoshas were. Although my knowledge of the culture and peoples of this diverse land was limited, it was difficult to imagine merging with any degree of success the concepts of tribalism and democracy.

10

A Strange Coincidence

The rumors started six months after Waldman took over the company with casual references dropped innocently enough in his conversation that usually went something like "Just think how much warmer it is in Florida now" or "We wouldn't have to shovel all this white stuff if we moved to Florida." At first, we paid no attention. A few warm days, and he would forget all about Florida. As the months passed, whenever Waldman talked about moving to Florida, he began prefacing it with *when* and not *if*. The trouble was, it made sense.

Many of our customers were Piper distributors, and most of the airplanes they took were built at one of Piper's two Florida plants: one in Vero Beach and the other in Lakeland. The Lock Haven plant was turning out only crop dusters and Super Cubs, which we never ferried, and Aztecs, which we ferried, but in decreasing numbers because Piper was phasing out that model. As disappointing as it was, logistics suggested Globe Aero should also be located in one of those southern cities. But the pilots fought it anyway.

The majority of Globe Aero's employees were from the northeast, many from Pennsylvania, where one got used to the luxury of four separate and distinct seasons. None of us could get excited about living in a hot, muggy climate with year-round sunshine. Even Donn Kerby, the lone southerner among us, was less than enthusiastic about a move that was beginning to sound less and less like one of Waldman's daydreams.

"I left Little Rock because it was too hot. Florida's even worse," Kerby griped, when several of us were sitting around the breakroom wondering about the likelihood of the rumored move. "Now why would anybody in their right mind want to move there?"

"But, Kerby, Phil told me just the other day that you couldn't wait to move back down south—so you could get some grits," said Paul Briggs, with an expression of exaggerated surprise.

The eldest of our group—and looking every bit the veteran, gray-haired, airline captain—Briggs liked to work at cross-purposes by getting people to disagree and become so wrapped up in debate that they forgot where he stood on the subject. As each of us offered our thoughts on the move, the discussion evolved into one of southern dishes and customs we would have to get used to until Kuney decided we had gotten off track.

"What the hell are we arguing about here?" Kuney demanded. "If we can quit all this gibberish, let me tell you how it's gonna be." The cacophony of voices quieted, and Kuney enlightened us as to his solution to the problem. "Phil can move to Florida if he wants, but I'm staying here. And that's all there is to it."

Kuney was an ex-Green Beret with a shaven head and intense eyes whose usually soft-spoken manner belied a commanding presence whenever he walked into a room. A modern-day mountain man, he generously consented to flying just to have something to do between hunting seasons. "Maybe I'll just fly down to Florida when Phil has a trip for me," he offered.

"But just think, Kuney. You can quit hunting and take up surfing. You might even become a beach bum," said Briggs, trying to goad Kuney out of his usually implacable manner.

"You don't seem to understand. The only thing they hunt in Florida is alligators, and I'm not going to move for that."

"I think we ought to get Phil to move to Boston," I said, unable to picture Kuney in a backdrop of surfboards and beach balls. "There's a lot more to do there. And it's got good airline connections, so we can get back from overseas a lot easier."

"Boston?" said Kerby, rolling his eyes and shaking his head. "What do we want to move to Boston for? You Yankees are always bragging about Boston. What in the world's so great about it anyway?"

"It's the hub of civilization. It's where man first swung down from the trees and became civilized. I think we ought to lobby Phil to move there," I said, though I had few supporters who considered Boston as a possible alternative and even fewer who accepted my theory of man's evolutionary beginning. Except for Dave Gray, my sole Yankee ally from Gloucester, just north of Boston.

"And it's a lot closer to Gander," said Gray. "We can clear customs right there and won't have to stop in Bangor."

"I like going to Bangor," said Kerby.

"But it's an extra stop, and it adds more time to the trip," I said.

"But the best thing about Boston," said Gray, "is that the weather in Massachusetts is better than Florida."

"What?" exclaimed Kerby. "You must be outta your mind. I never heard o' such nonsense."

"It's true," said Gray. "It rains almost every day in Florida—in the summer, anyway. Just ask Bennett. He'll tell you. Thunderstorms every day. You can set your watch by them. You can't argue with statistics, Kerby. If you look at it over a year's time, Massachusetts has more sunny days than Florida."

"It sounds like you might be reaching just a little," I said to Gray. "But did you mention that to Phil?"

"Yeah, but he just laughed and kept talking about how great Florida was."

The pros and cons of the move were debated for months until Waldman made all further speculation academic when he told us he was having a hangar built in Lakeland and Globe Aero would be moving by next spring. When we unanimously complained, Waldman continued to give the usual responses concerning snow and sunshine, as well as the classic, "Just wait'll you see how much money you save because you won't have to buy winter clothes."

"You're kidding," Rice said, as we brushed the snow from our parkas and walked into the Albatross Hotel. "He didn't really say that, did he?"

"Doesn't that sound like something Phil would say?"

"I guess it does," Rice agreed, "since he doesn't take trips anymore. He doesn't have to walk across an ice-cold ramp or shovel snow off his airplane. He's home sleeping in his own bed every night, and the few times he does go overseas to wine and dine customers, he's in the passenger cabin of a Boeing 747, where the only ice he has to worry about is in his martini."

"I think I was home six days last month," I said, "when I wasn't puddling across the Pacific."

"Well, you're not even doing that now. Doomsday says we're going to be here until this blizzard moves, and he didn't even want to guess when that would happen."

As usual, Rice and I were holed up in the Albatross Hotel bar, watching the snow fall and wondering what to do next. Gander wasn't exactly a thriving metropolis; there were several bars and restaurants, two malls, one movie theater, and a population of eleven thousand, many of whom seemed to be involved in one way or another with aviation. Unlike the rest of Newfoundland, where coastal villages depended on fishing for their livelihood, Gander's growth, and its very birth, was due to its distinction as the closest North American city to Europe.

Gander's humble start was as a wilderness area known as Hattie's Camp at mile 213 on the Newfoundland Railway. It began to grow in the late 1930s, when it was decided an airport was needed for making experimental flights across the Atlantic. The governments that commissioned the airport project, however, didn't take into consideration that nine months out of the year Newfoundland was under the grip of Arctic weather and three months by mosquitoes. Construction progressed slowly at Gander, named, according to local legend, because another airport was being built in Goose Bay and the military, which was overseeing both airport projects, wanted a Gander for the Goose.

With the end of the flying-boat era in the early days of World War Two, the airport was built up by the U.S. Air Corps to ferry bombers to Europe. When the war ended, Gander became a major staging point for transatlantic traffic. The small village grew, and hotels began sprouting up to accommodate the passengers and flight crews. On any given day, in any of its four hotels, the lobbies were crowded with an assortment of airline crews in various stages of their nomadic profession: a few uniforms checking in at the registration desk; several waiting for taxis; while others already in civvies were scattered throughout the restaurant and lounge. There were Britian's Royal Air Force and German Air Force pilots, the Air Cubana crews, the proper English crews of BOAC, the secretive East Germans, and, almost always, one or two ferry pilots.

"You know, we spend so much time here, maybe we should move to Gander," Rice suggested.

"Yeah, that's a great idea," I said. "We're bored now after only two days. If we moved to Gander we could be bored a whole ten or fifteen days every month."

"Gander's not such a bad place. It's a growing city, and you can get land cheap. We could build an A-frame right on Gander Lake."

"So we build an A-frame, then what? There's not much going on in Gander now; we could build a hundred A-frames on the lake and it wouldn't change a thing."

"But this one'll be different," he said, smiling as he let me in on his great idea. "We can turn it into a bordello."

That was what I liked about Rice—he was always cooking up some crazy scheme. "That certainly ought to liven things up around here," I said. "But I don't think that's what the town fathers had in mind when they founded Gander."

Rice had it all worked out and eagerly suggested that we could avoid corrupting local morals by catering strictly to airline crews that, like us, were probably tired of going to the same bars and eating in the same restaurants. "We have to come up with something different, though. Something unique, kind of exclusive . . . to keep out all the riff-raff."

Rice's enthusiasm was contagious, for now I too was getting hooked on the idea. I envisioned a piano bar where patrons could join in a boisterous sing-along with a handlebar-mustached, straw-hatted piano player before retiring to a private room for more carnal pleasures.

"And I've already thought up a name for the place," Rice stated, with obvious pride. "We can call it the Chicken Ranch."

I thought of that famous establishment in Nevada with the same name and then had an inspiration. "I've got a better idea. We can call it the Gander Chicken Ranch; because if it goes over big here, we can open them in other cities." Our empire grew to include the Bangor Chicken Ranch, the Honolulu Chicken Ranch, and the Golden Gate Chicken Ranch.

"We can start up a whole chain of these places and retire from ferrying," Rice said, warming to my ideas of expansion. "Oscar Lima would like that."

"How do you think she'll like it when you tell her you're going to be a pimp?"

"You mean a gentleman pimp."

Before we could implement these plans, however, something happened that we hadn't expected for a few days yet. The weather cleared, and we had to continue with our trip.

Flying always seemed to get in the way. It curtailed our plans of becoming rich carpetbaggers, and it did nothing to abet our romantic endeavors. On those occasions when the gods smiled on us and we ended the evening with some enchanting young lady, our ecstasy often ended the following morning when we whispered into our partner's ear that we had to "check with the weather office." It was an endearment that undoubtedly left her wondering what the weather had to do with postsexual snuggling.

As inappropriate as it might be at the time, checking in with the weather office was part of our daily routine. Every morning, rain or shine, drunk or sober, we had to check in with the weather office as if it were some pagan ritual to appease the gods. Not satisfied with hearing the bad news over the phone, we would often go to the airport and stare at the weather charts as if willpower alone would make them change. No matter how hard we stared, rarely were we successful. What we usually ended up doing was trying to coax a more promising forecast from one of the briefers. Although we probably made a nuisance of ourselves with our repeated visits, the briefers never let it show. More than most people, they understood the limitations of our aircraft and gladly spent extra time analyzing the movement of a meandering low-pressure system and the possibilities of icing.

Meetings between pilots and briefers are usually short and quick. We exchange a few comments about the weather and then part, seldom seeing each other again. In Gander, because we spent so much time here, eventually a close bond was formed. Ron Mills, Clayton Jeans, and the others had been briefing pilots on North Atlantic weather since World War Two and knew the ocean better than most people knew their own backyard. To us, they were the high priests of the North Atlantic. We flew when they told us to fly, and we stayed on the ground when they advised that to be the best course of action. When we got in trouble they worried about us, and when one of us went down, they mourned the loss of a friend.

The focal point in the Gander Met Office was a large map on the counter that depicted the weather over the ocean. It was redrawn every six hours to reflect the latest data and then placed in a leather desk pad. It looked like any chart one might find in any weather office. On the underside of this chart, however, handwritten on the cardboard backing, were the names of the pilots who had left Gander over the years and the dates when they went down over the ocean.

Ditchings were an almost inevitable part of ferrying. Too many pilots ditched because they picked up too much ice, their engine quit, they collided with another aircraft, or they ran out of fuel. George Martin, one of the old timers in the business, went down when he encountered unforecast sixty-knot headwinds and heavy icing off the coast of Iceland. He died struggling in the icy waters trying to get aboard his life raft. Ironically, had he stayed with the airplane he probably would have survived because it floated for three days.

If there was one thing I didn't need to be reminded of before a crossing, it was that my name could easily end up with those of all the other dead pilots. For some reason, I had gotten into the habit of occasionally reading the names to see if I recognized anyone. Other than Martin, rarely did I come across a name that even looked familiar, except for the one time I not only saw the name of a pilot I knew, but a good friend: "Dave Gray 3/19/75."

Of all the entries, I knew this one to be an error, because the simple fact was that Gray was not dead. Gray himself had told me about it, as well as the strange set of coincidences that occurred on that night over the Atlantic.

It had all started when Gray left Gander in a Piper Seneca heading for Shannon and was several hundred miles out when the alternator on his right engine quit. Because the alternator on his left engine was still working, he kept going toward Shannon. One hour later, when his left engine quit, Gray knew the situation was serious. In another hour his battery would be dead and his radios useless. Although the airplane was light enough to maintain altitude on one engine, it required almost full power to do so. With the airplane flying much slower, Gray estimated he would run out of fuel before reaching Shannon.

The moment every ferry pilot dreaded had arrived—Gray would have to ditch. Of that he was certain. The only questions were where and when. Either fly as long as he could and hope the airplane didn't run out of fuel until he was near Shannon or ditch close to Ocean Station Juliet while he still had power and could control the ditching. The latter seemed the better choice because he wanted to be as close to the rescue vessel as possible when he hit the water.

Gray was in his early twenties and had known he wanted to fly for the airlines since he was a teenager in his high school aviation club. He had learned to fly at Embry-Riddle University and then took a job fish spotting for the fishing boats in his hometown of Gloucester. He had little flying time, but before the night was over, he would have more experience than he had gotten in his entire flying career.

He was still three hours away when he relayed an SOS to a BOAC airliner, asking them to advise the ship of his intentions. A short while later his battery died and the aircraft went dark. Luckily, it was a clear night, so he was able to keep the wings level and the aircraft pointed in the same general direction. Not only was he unable to talk to other aircraft or the ship, but his navigation radios were useless. All he could do now was rely on time and distance to find the ship and hope the winds wouldn't blow him off course.

One hour later he passed a few miles south of the ship, lit up in the black ocean like a small city with a long, straight line of flares swaying in the water and outlining a makeshift runway. To anyone piloting a small, disabled airplane over the ocean at night, there could not be a more welcome sight. But for Gray, it was conceivably the happiest moment of his life, not just because he had found the ship, but because he was over an hour early. If the winds held up and if his fuel calculations were right, he figured he might be able to make it to Shannon.

With no radio, there was no way he could call the ship and let them know of his change in plans. They would be looking for him, waiting to pluck him out of the sea, but all he could do was call them when he landed. Crabbing into the wind to compensate for a crosswind he suspected was stronger than forecast, he maintained his heading until he estimated he should be over Shannon. It was an hour before sunrise, still dark, but light enough that he could see a blanket of fog covering the earth. Unable to determine his position, he turned north, hoping the colder temperature would dissipate the fog and he would be able to find an airport before running out of fuel.

Twenty minutes later he spiraled down through a hole in the clouds, reaching their bases a few hundred feet above the ground. Although darker, and with clouds blocking the sun, Gray could see that he was in a valley surrounded by hills whose peaks were obscured by the low-hanging clouds. He quickly climbed back up and headed north again.

The fuel gauges were almost on empty when he found another small hole in the clouds and banked the airplane into a steep descending spiral. Tightening the circle to stay out of the clouds, Gray dropped lower until he was underneath the cloud layer a thousand feet above the flat ground. He began flying in a random direction, searching in the dim, overcast dawn for an airport.

He flew over rolling hills, small pastures, and thick woods, but no airport. After fifteen minutes he gave up the search for an airport or a large field and settled instead on a narrow road that ran straight for about fifteen hundred feet with a bridge at one end and a bend at the other end that faded into darkness. He began a slow glide, aiming for just past the bridge as his touchdown point, banked the airplane onto a short final approach—his wheels just clearing the bridge—and settled onto the macadam.

It was a timeless country lane with a few lonely houses on either side whose pastoral quiet was interrupted by the sound of Gray's airplane taxiing down the rutted pavement as if he were out for a leisurely Sunday morning drive. He pulled into a driveway and parked the airplane, walked up to the house, and knocked on the door. No one answered, so he tried the house next door. But again, nothing. As he walked back to the first house, he heard a distinct "clop, clop, clop" coming from behind him. Turning to see who or what was making the noise, he was greeted by the sight of a pure white horse skittishly prancing toward him.

Glad for the company, Gray got some fruit from his airplane and shared it with his new friend. After several minutes, a lorry chugged around the bend in the road. Gray flagged down the mud-smeared truck and asked the driver where he was.

At first the driver just sat there with a puzzled look on his face, his eyes going back and forth between man, horse, and machine. "Is that your aeroplane?" he timidly asked.

"Yes," said Gray. "I just made an emergency landing. Can you tell me where I am?"

"But . . . what're you doing with Tom Duff's horse?"

"The horse? Oh, he just followed me."

"Followed you? This sounds awful suspicious."

"Look, I had a problem with the airplane, and now I don't know where I am. Can you call the police and tell them I made a forced landing and ask them to send someone out?"

"Well . . . okay. But I don't know about this," said the lorry driver, before driving off and leaving Gray to the company of his four-legged friend. After waiting a few hours, he was about to try one of the houses again when he heard the sound of diesel engines and then saw two armed personnel carriers emerging from the bridge. Close behind the second APC was a tank and what looked like an entire squad of soldiers advancing across the field with rifles drawn. They pulled up right in front of Gray and stopped.

"What are you doing here?" a young captain demanded, as he jumped down from the vehicle, his firearm pointed at Gray.

By the insignia on his uniform, Gray knew that the soldier was British Army. He also knew there was only one place in Ireland with this many British soldiers—Northern Ireland. "Don't shoot. I'm a friend," he said, raising his hands as he proceeded to tell the soldier about the engine and electrical failure and how he ended up landing here.

"Do you have some identification?" asked the officer.

"I've got all kinds of identification. Here, let me show you," Gray said, as he reached into his pocket. There immediately followed several clicking sounds. Metal on metal. Gray's hand stopped just short of his pocket and he looked at the officer.

"Go ahead. But take it very slow," the captain warned.

Gray took out his wallet, slowly, smiling at the grim-faced soldiers as he took out his passport, pilot's license, driver's license, library card, and anything else that might convince the officer he was not an Irish Republican Army terrorist.

After studying each document and being reasonably convinced that Gray was who and what he claimed to be, the captain told him why they weren't taking any chances. "The IRA set up an ambush at this bridge just two days ago, and two policemen were killed. So when the police were told by a lorry driver that something strange was going on near the bridge, they notified us. We thought there was going to be another ambush."

As they discussed Gray's predicament, an old man came out of the house. His front yard was now a circus of activity. "What's going on out here?" he asked.

"Everything is okay, sir," the captain said. "This man just made an emergency landing."

"Well, now, I guess we have to figure out how to get you on your way," said the man. "Why don't you come in and have some breakfast, and we'll see what we can do. That's all right with you, Captain, is it?"

"Yes, sir. We have to watch the airplane while it's here," the officer said, and then turned to Gray. "We'll stay until you can find a mechanic and get going again."

Gray followed the man into the house and sat down at the kitchen table, while his host rummaged through cupboards and rambled on about how Old Duff's horse had figured out how to unlatch its stall and roamed around the neighborhood whenever it pleased.

"Hope I didn't wake everybody up," Gray said, as the old man set two plates of bacon and eggs on the table. "Where am I anyway?"

"Well, now, the thing is, you just happened to land on the Main Strabane to Londonderry in County Tyrone. I guess you must be going to Shannon. You didn't miss it by much—three hundred kilometers maybe. You watch these eggs now, whilst I see about getting someone to fix your aeroplane."

"Don't tell me you've got an airplane mechanic here?"

"Aeroplane? Now, we don't have no use for an aeroplane mechanic. But we got a good electrician; works on my lorry all the time."

Three hours later, the electrician had the engine running and had hot-wired the electrical system so the alternator wouldn't quit again. With the soldiers' help, Gray siphoned thirty gallons of fuel from one of the army trucks, thanked the helpful Irishman and the soldiers, and took off for Shannon.

As far as Gray was concerned, the adventure was over. The fog had lifted, both engines were running, and he was once again over safe, familiar ground. All was not as calm on the opposite side of the Atlantic, however. The last message air traffic control had received was that Gray planned to ditch at Ocean Station Juliet. But the time he should have arrived there had come and gone, and no one had heard from him. He had to have used up all his fuel by now. If not already, Gander Center would soon be suspecting the worst.

Meanwhile, in the North Atlantic, not far from where Gray had been nursing his sick airplane, a ship spotted the wrecked tail section of an aircraft floating in the water and radioed the registration numbers on the tail to Gander Center, where they were cross-checked against flight plans for aircraft presently over the ocean. When the numbers matched those on Gray's airplane—the same airplane that declared an emergency because of an engine failure and then didn't show up at Ocean Station Juliet—air traffic control notified the weather office that the airplane had ditched, broke up on impact, and the pilot was presumed dead.

The strange thing about this incident was that the wreckage was part of an airplane that ditched several months earlier. When it was confirmed the aircraft went down over the ocean, the Federal Aviation Administration—as was the routine—canceled the registration and assigned the number to the next aircraft to come off the production line. By sheer coincidence, the aircraft to get that registration number was the one Gray was flying. This was unknown to the controllers and briefers in Gander. As far as they knew, Gray was dead. So, they entered his name with the other ghosts on the underside of the weather chart.

Gray, however, didn't stay dead. A few weeks later he walked into the weather office and was greeted by a shocked Clayton Jeans.

"Gray!" exclaimed the usually jovial briefer, when he looked up and saw the pilot standing at the counter. "Is that really you?"

"What's the matter, Clayton? You look like you just saw a ghost."

"Well, uh . . . you're supposed to be dead," he stammered.

"Dead? Where'd you hear that?"

"From ATC," said Jeans, no longer sure of what to believe. "They told us you lost an engine and went down. They said they found the wreckage of your airplane."

"Oh, that was all a mix-up. Look, here I am. Real," Gray said, grinning at the confusion his recent adventure had caused and offering his hand as proof of his claim.

Jeans responded slowly, as if expecting the apparition to vanish before him. All that changed when he shook Gray's hand and felt real flesh. The worried look gone, he began smiling, laughing, and hugging Gray as if he were a long-lost nephew. Then, all of a sudden, he stopped; the smile left his face, he stepped back, and cautiously said, "Now you're not going to like this, but I'm afraid it's already done."

"What're you talking about? What's done?" Gray asked.

Jeans lifted the weather chart on the counter to show the underside. Recently added to the list of pilots who went down over the ocean was Gray's name and the date of his presumed death.

"I didn't put it there. Honest. Let me get rid of it," Jeans said, apologetically. He was about to erase all traces of Gray's demise with a black magic marker when Gray stopped him.

In any type of work where life and death depended to a large degree on luck, it was not uncommon for superstition to be so highly regarded. In keeping with this belief, each of us had our earthly symbols of paganism that we clung to as if our very lives depended on them: an old life raft that was probably mildewed with holes and would burst apart if someone were foolish enough to inflate it or an emergency transmitter whose batteries had long ago become useless with age. But we kept them anyway. These were the rabbit's feet and four-leaf clovers of our profession. We had survived this long with them that their presence alone suggested we would have no use for them.

Gray was no more superstitious than the rest of us. Yet, staring at his name affixed to this solemn testament to those pilots no longer alive seemed the perfect talisman against something he could not control. "I didn't die," he said to Jeans, "so it's probably good luck. Why don't you let it stay."

11

The Making of a Guru

There were the love bugs. Alligators cruised the streets and caused traffic jams. There were sink holes big enough to swallow a Greyhound bus. And just as Gray had said, there were thunderstorms every afternoon. It was April, the time of year winter residents left Florida for their annual migration back to permanent homes up north, when Globe Aero moved to Lakeland. Although we griped about it, everyone followed Waldman except Wheeland and Rice, whose career priorities and spousal objections precluded such a move, and Kuney, who refused to leave his precious mountains of Pennsylvania.

Acclimation to this new climate, in spite of our reservations, was not without its benefits. For one thing, we moved into a new hangar built specifically for our type of operation. It had offices in the middle with open bays on either side for working on airplanes. Waldman was especially proud of this design, claiming it would speed up the entire tanking process.

"You just taxi in one end, Probst'll throw some tanks in, and you can taxi out the other end. You won't even have to get out of the airplane."

Another good thing about Florida was that everybody moved to the same city. In Pennsylvania we were so spread out in neighboring towns that the only time we saw each other was when we were on a trip or at the hangar preparing for a trip. Now, with everyone living so close, we began to get together socially as well as professionally. Being new to Lakeland we naturally gravitated toward our only focal point—the new hangar. It was a place to hang out when not on a trip and be with other people we shared a common bond with. There was always at least one pilot in the office, and more often than not that pilot was Kerby.

From the beginning, Kerby was different. His background offered no indication he would become an expert at ferrying airplanes. Before coming to work for Globe Aero he was earning a sketchy living as a crop duster, a job that, in his own words, did not have a lot of long-term prospects.

"Most crop dusters end their career wrapped around a telephone pole," he said, when people asked what motivated him to make a change in livelihood.

When the spraying season ended, he ferried agricultural airplanes from Lock Haven to El Paso. Although more interesting than crop dusting, it was still boring work that didn't pay much. Some friends had told him about a hotel in Nicaragua where all the mercenaries and pilots stayed, and that if he wanted to make some real money that's where he should go. Heeding their advice, Kerby delivered his last airplane and was all set to leave for Managua when he got robbed in a whorehouse in Ciudad Juárez in Mexico. One week later he was back ferrying crop dusters and running into people who kept asking him if he worked for Globe Aero. Kerby decided then, that if he was going to ferry airplanes, he ought to look into working for a company that flew to more interesting places.

That was two years ago. Globe Aero was still in Lock Haven, and Walt Moody was still the owner when Kerby visited the hangar.

"So you're ferrying that crop duster," Moody said. "Where are you taking it?"

"Just down to Fort Lauderdale," Kerby answered.

"Since you're going to Florida, anyway, how about going to Vero Beach and pick up an airplane for me and bring it back to Lock Haven?"

Kerby delivered the crop duster, picked up a brand new Piper Cherokee, and knew right then that this was the job he had to have. The airplane had a full set of radios and an autopilot—equipment entirely foreign to crop dusters. When he brought it to Lock Haven, Moody was so impressed with how fast he got back that he hired him on the spot. Kerby ferried airplanes around the country for six months before Moody sent him to Australia with Rice and Weir. They spent twenty-three days in San Francisco waiting for the winds to improve over the Pacific.

A few trips later he and Rice were again on their way to Australia flying what, at the time, were the smallest aircraft we ferried in the Pacific—two single-engine Aero Commander 112As. After a twenty-hour flight to Honolulu, they left the next night for Pago Pago, but had to return because of headwinds. Kuney, also in Honolulu waiting for the winds to improve so he could continue on to Australia, did not think this did much for their professionalism.

"What the hell are you guys doing?" he demanded. "You're going off half-assed like you can't even plan a simple flight. It makes us all look unprofessional. If I was chief pilot I'd fire both you guys."

Kuney was as disciplined in flying as he had been in combat in Vietnam. He was a little bigger than Kerby, an easy-going but uncompromising southerner, at times too staunch in his beliefs to back down when good sense argued otherwise. Kerby listened to Kuney's tirade until he decided he had heard enough and, in an accent as thick as a hot Arkansas afternoon, told him that if he didn't like it he could go blow it out his ear.

"You wanna straighten this out right now?" Kuney suggested, as he stood, threatening Kerby to respond.

Kerby's fists were clenched in his pockets and his neck was beginning to twitch, as he, too, stood and matched Kuney's bravado. "Okay, let's go," he said.

After trading insults and threats, and glaring at each other like pit bulls, Kuney decided not to let this go any further because he didn't want Kerby to bleed all over him. They got along great after that. Kerby stayed with Globe Aero and, after overcoming the stigma of being the new guy with the funny way of talking, we began to trust him and quit mimicking his accent.

Kerby and I hadn't flown a trip together and didn't know each other that well when I ran into him at the hangar in Lakeland after he had just returned from Singapore. I was about to give the usual greeting, but in light of his condition, decided something else would be more appropriate.

"Jesus, Kerby, you look like hell! What did you forget to duck?" Scratches covered the right side of his face, he favored one leg, and when he looked at me through bloodshot eyes behind lopsided glasses he seemed to be staring off into space. Usually more animated with movement and gesture, his every move seemed to require great effort.

"You would, too, if you all went through what I just did. I didn't think I was gonna make it back alive."

Suppressing the urge to laugh at his comical appearance, I suggested he tell me all about it over dinner. Dining with Kerby was not merely a function of stuffing food into one's gullet. To him it was an occasion to nourish all the senses and observe all the rituals of fine dining. Benny's Oyster Bar was not the type of place that lent itself to that type of ritual, but it had good seafood and it was quick. Kerby ordered the largest lobster, along with a generous serving of raw oysters and, after expressing his disappointment upon learning that there was no wine list, made the best of their limited selection before beginning his tale.

"I was over New Guinea when Center strongly suggested I land at Rabaul because I didn't have the right HF frequencies and wouldn't be able to talk with them when I got out of VHF range. So I landed, and then the controller tells me I can't leave because I don't have the right frequencies. 'You're a hazard to navigation,' he says. Imagine that! Me, a hazard to navigation," Kerby said, laughing. "But now that I was on the ground, they decided that as a health precaution I should be given inoculations for every conceivable disease: yellow fever, cholera, typhoid, and several others I never heard of."

"Don't you ever have a normal trip?" I asked.

"I keep trying," he said, and tasted the wine that the waitress brought to our table. Finding it satisfactory, he nodded to the waitress before continuing. "After they gave me all these shots, I got sick as hell and spent the next two days in bed. I felt better the third day, so I made arrangements to relay position reports through an airliner and left for Wewak. But when I landed, five Australians were waiting for me. I was still having radio problems, so I didn't know if they were gonna fine me or throw me in jail or what. It turned out they were just curious about someone flying a small airplane from the United States to Singapore and wanted to meet the crazy American that was doing it. They told me I could relay my position reports with another airliner for the next leg, but that it wasn't scheduled for another two days. But since I had to stay there anyway, they asked me to give a speech at their Rotary Club."

"That's a new one," I said, chuckling. "Nobody's ever had to give a speech before. I hope you didn't set a precedent. What'd they want you to talk about anyway?"

"Well, uh . . . ferrying airplanes, I guess. I didn't know what to talk about. But the way it was put to me by these Australians, it sounded like my departure was contingent upon giving the speech. So I told them I didn't see how I could refuse."

"That was probably a wise move. So what'd you talk about?" I asked, as the waitress placed a small plate of fish in front of me and a heaping pile of lobster in front of Kerby, who, with the measured precision of a surgeon, began deftly separating meat from shell.

"I just told 'em a bunch of stuff about ferrying airplanes around the world. You know, the long distances, the typhoons, and picking up ice in the Atlantic. They were fascinated with everything and kept asking questions. 'What was my favorite country?' and 'Did I ever have any close calls?' and 'What type of airplane was I flying?' "

"I would've liked to have been there when you told them you were flying a little Piper."

"Well, uh," he said, laughing, "I tried to tell them, but they didn't believe it could be done. 'A small airplane with only one engine? Over all that water? You can't do that.' Jesus! They all thought I was gonna die. So everybody wanted to come out to the airport the next day to see me take off for themselves."

"They must've thought you were a hero when they saw you climb into this rinky-dink little airplane."

"But they didn't see it. I went to the airport and hung around for a while, but nobody showed up. They probably thought I was gonna kill myself, and nobody wanted to be the last person to see me alive. Probably figured it was bad luck or something. So I left, I guess around mid-afternoon and flew all night."

"That's not exactly the best time to be flying in that part of the world, you know. They get some pretty hellacious storms down there."

"Yeah . . . well . . . I know it now. I ran into a lightning storm and was in solid clouds for about seven hours. I was flying along the equator, over New Guinea and Borneo, and it just kicked the hell outta me. Course, wouldn't you know it, now I had the right frequencies, but I couldn't get out on HF because of all the static. So I didn't talk to anybody 'til I landed at Kuching just after sunrise. The next morning I just wanted to get the hell outta there and deliver the airplane, but now I couldn't leave because I had a dead battery and a bad case of diarrhea."

"With the things you eat, somehow that doesn't surprise me."

Kerby sucked the last trace of meat from the lobster's claw before responding, "You all gotta experiment some more. You should try some of this lobster. It's actually not bad."

I took him at his word and bit into my halibut. Never having acquired the taste for lobster myself, I nonetheless enjoyed watching other people wrestle with them while turning the table into a junkyard of used napkins and discarded shells. A general rule of thumb I had learned about eating lobster was the bigger the mess, the greater the enjoyment. Based on this crude measurement, I gathered Kerby was having one hell of a good time. "So how'd you get out? Kuching isn't exactly geared up to do maintenance on small airplanes. I presume you did get the airplane fixed."

"Yeah, but it took a while," he said, while wiping a trickle of lemon juice from his chin. "I spent about an hour in the airport restroom to take care of the diarrhea. Then I got two mechanics and bought them a case of beer to hand-prop the airplane, and took off before anything else went wrong. But now it was getting late, so I didn't get to Singapore 'til after dark. Of course, everyone had already gone home except for that Indian Pat White's got working for him. He was probably pissed off because he had to stick around and wait for me, so he just handed me a flashlight and told me I'd have to detank the airplane myself."

"He didn't even help?"

"Nope. He just went back into the office and left me to do it all myself. So I got a flashlight in one hand and a wrench in the other, and I'm pulling the tanks and all the fittings out. There was stuff all over the ramp. Then I started cramming everything into four big suitcases. By the time I got to the hotel, it was almost eleven o'clock and the restaurant was closed. So I asked the concierge if anything was open, and he tells me there's an open-air food market a few blocks away and if I hurry I might be able to get there before it closes. So I threw my bags into the room and went back downstairs. I went out the back door of the hotel. It was dark and all because they didn't have any lights on, and I got about fifty feet before I fell into a drainage ditch."

"C'mon, Kerby, nobody falls in those ditches," I said. "You sure Rice wasn't out there somewhere, sending out vibes?"

"I'm not kidding. I cut the hell out of myself. I had a big gash on my leg, my pants were all torn, and when I tried to stand up," he said, grimacing at the memory. "Jesus, I never felt so much pain. I finally pulled myself up, though, and tried to get outta the ditch, but it was too deep. I was standing on my toes, reaching for the top of the sides, and all this pain was shooting right up my leg and I just couldn't get a hold of anything. So I looked for a low spot or something to stand on and started yelling for help."

"You're lucky it wasn't monsoon season; those ditches are probably flooded with water then," I said.

"Oh, Jesus!" he said, rolling his eyes toward the ceiling as he made a cross over his chest. "I'd have drowned for sure. Anyway, I kept walking along the ditch and getting worried because now it didn't look like there was any way out. I must've yelled for two hours before some cop heard me and pulled me out. Then he wanted to know what I was doing in the ditch."

"I'm surprised he didn't just throw you in jail. They don't have much of a sense of humor over there. I guess you were glad to get that trip over with."

"Oh, it wasn't over yet. I still needed to get something to eat. It was one-thirty in the morning when I got to the street market. It was still open, but most of the push carts had been cleaned out and all I could find was a large pan of small, raw clams."

"You didn't eat raw clams? What're you nuts?"

"That's all they had left. Besides, I was tired and sick from all the shots I got in Rabaul and the diarrhea in Kuching. I was limping because of the pain, and, most of all, I was starving. So I ate the whole bowl of clams. But soon as I finished, my stomach started growling. Great, I thought, on top of all this, now I get food poisoning."

"You know, our ancestors spent a few million years learning how to make fire so they could cook their food. It'd be a shame if we didn't take advantage of that discovery."

"Well, I had to eat something. But let me finish."

"You mean there's more?"

"When I got back to the hotel I went straight to bed. But now I couldn't fall asleep, because all these hookers kept knocking on my door. I guess the concierge kept sending them over. Either that or they were doing some free-lancing. When I finally did fall asleep, I didn't wake up until seven that night—twelve hours after my flight left. So the next morning, I went to the airline's downtown office to get my ticket rewritten. But when I got back to the hotel, I couldn't find the ticket and spent the next two days looking for it. I finally found it and left the next day before anything else could go wrong."

The problems Kerby ran into on this trip turned out not to be atypical of the many he flew for Globe Aero. No matter where he went, something happened to him, which he laughed about later and figured it was just part of the adventure. He may have started out like the rest of us, fumbling his way around the world not always sure of where he was or if he should even be there, but he caught on quickly. Interestingly enough, Kerby actually liked flying twenty hours to get to some remote island that none of us had been to before. "Kota who?" he said, after trying unsuccessfully to repeat the name when someone told him about Kota Kinabalou. Kerby seemed to have a simple formula: if a place was hard to pronounce, that was where he wanted to go. Kerby had a deep southern accent, and there were many places he had trouble pronouncing.

Although he had been flying for Globe Aero only a short time, already he was becoming somewhat famous, especially in the Pacific. It seemed as if every place I went to, someone either knew Kerby personally or had at least heard of him. I began wondering if there was no place where he hadn't made friends. I thought I had finally found one when I ferried an airplane to the government of New Caledonia, an island group near Australia, nowhere close to any of our normal routes and, to my knowledge, no Globe Aero pilot had ever been to.

Here was a place where nobody could possibly have heard of Kerby, I thought, when I landed at Noumea and shut the engines down in front of a large gathering of people. Exiting the airplane, I walked over to the high commissioner who was there to officially welcome me to New Caledonia. As we shook hands and smiled for the cameras, the first thing he asked was, "And how is my friend, Donn Kerby?"

Kerby was becoming a guru for all things having to do with ferrying, for there was almost nothing anyone could ask him and not be given an answer. If not right away, then a few days later, when he produced a thick file folder complete with facts, figures, and several magazine articles on the subject. If anyone wanted to know what the winds were like over the Atlantic for any given month, Kerby found it in an old weather folder. If someone wanted to know what restaurant to go to in a particular city, Kerby not only came up with a name but a description of food and ambiance so exotic one might think they were talking to the restaurant critic for *National Geographic*.

While most of Globe Aero's pilots had a small notebook with information on some of the places we flew to, they consisted mostly of airport and weather-station phone numbers and performance data on airplanes. Kerby, however, had several notebooks, every one of them filled with train and airline schedules, phone numbers of nightclubs, hotels, restaurants, contacts that could help if we got into trouble, and anything else he considered to be of more than passing interest. If any of us were about to leave for some place we had never been to before, all we had to do was ask Kerby. But we had to be prepared to take notes.

"When you land," he would begin, "be sure to tell the tower you wanna park on the south ramp, because that's where they're gonna refuel you. If you don't say anything, they'll send you over to the boondocks, and then you gotta walk half a mile to the airport office to tell them you want fuel. Then you gotta walk back to the airplane and taxi over to the south ramp, which is where you wanna be in the first place because it's close to the met office and the terminal. After you finish refueling, get one of the girls in the traffic office to make hotel reservations for you. Try to get a room at the Palace. It's one of the better hotels, and they give you a crew rate. Just ask for Harry, the head desk clerk. He's a buddy of mine, and he'll take care of you if you tell him you fly for Globe Aero. He'll wanna take you out and get you drunk and then go to a snake shop for dinner. Now everybody ought to try that at least once. They put a snake in the same cage as a mongoose, and after they get finished fighting it out, they'll cook the winner for you. And don't forget to bring some Oreo cookies, because they can't get any there."

Kerby's fanciful tales of islands that until now had been little more than quick refueling stops or a place to grab a few hours sleep were contagious. We began to appreciate not only the logistics of our work, but also the different cultures we came in contact with. Though we weren't consciously aware of it, a change was taking place in the way we viewed our work. No longer were we scanning the airway charts to find the most direct routes, now we were looking for new islands to explore. More adventurer than pilot, we likened ourselves to Ferdinand Magellan, Sir Walter Raleigh, James Cook, and all the other explorers of earlier centuries who stuck a flag in the soil and claimed all for England or Spain.

In the late twentieth century, however, people didn't go around planting flags in the ground, laying claim for the country that financed their expedition. For one thing, the people living there might take a dim view of it, and for another, we didn't have any flags, baggage stickers, or anything proclaiming we had visited a certain place. Other aviation outfits had them—airlines, military squadrons, air survey, and rescue crews. We couldn't go into an airport hotel bar or weather office without seeing at least one entire wall covered with a collage of tigers, shamrocks, globes, and every other conceivable logo. Yet, there was not one for ferry companies.

As Globe Aero's resident artist, I felt it my responsibility to design a logo. I came up with an airplane originating in Florida and circling the globe. So that it stood out from all the other baggage stickers and also profess a discerning worldly appreciation for the finer things in life, I added a legend around the border that proclaimed: THIS ESTABLISHMENT RECOMMENDED BY THE PILOTS OF GLOBE AERO.

I had a few thousand of them printed and then gave them out to the pilots. Before long, they started showing up in airports, hotel lobbies, restaurants, bars, train stations, and anywhere else we felt like sticking them. I left one at the ski lodge on Mount Buller, Australia, Briggs stuck one at the Marina in Singapore Harbor, and the contest to see who could place one at the most remote, out-of-the-way place had begun. Flying to new places, though, no matter how adventurous, was not without its share of problems. We might go several months when every trip went off exactly as planned and start thinking nothing could go wrong, but then we would have a run of bad luck and no matter what we did, nothing went right.

Tim Peltz and Bruce Copp were halfway to Christmas Island in the southeast pacific when they found out that no one turned on the radio beacon. They spent the next several hours searching for a tiny smudge of coral and sand, while their aircraft guzzled gallon after gallon of aviation fuel. Although they eventually found the island—after twenty-five hours aloft—Copp had only a small amount of fuel left and Peltz ran out as he touched down on the runway.

Then Roger Dowst flew to Lord Howe, a small outcropping of rock four hundred miles east of Sydney, Australia. The problem with Lord Howe was that the runway was sandwiched between two rock formations that acted like a wind tunnel, which made landing in even the slightest breeze an interesting show for any spectator. After Dowst blew two tires while landing in a crosswind and had to wait four days for new ones to be flown in from Sydney, we decided Lord Howe was an island best viewed from the air.

Of all the islands we flew to in the Pacific our favorite was Norfolk Island, a tiny oasis of green halfway between the Fiji Islands and Australia. Approaching from the air, one is immediately surprised at the sight of dense green forests of towering pine trees that seemed more like the state of Maine or the Pacific Northwest. This was not what I expected to see in the middle of the South Pacific, where long sandy beaches, coconuts, and palm trees awkwardly bent under the stiff breeze of an approaching hurricane tended to be the norm.

Just a few miles from the airport was the small pier and the old prison that once housed mutineers from the HMS *Bounty*. It was a sad testament to the United Kingdom that during the centuries when the tentacles of their empire crisscrossed the globe and the East India Company roamed the seas at will, they required a vast network of penal colonies.

With a rented motorbike from the small village of Kingston, I rode into the island's interior along narrow country roads that wound through secluded valleys and up steep hillsides, past lonely farmhouses and an old cemetery with wild flowers and weather-beaten tombstones dating back to the seventeenth century.

Reading the gravestones, I followed a genealogical trail left over the centuries: "Elizabeth, beloved wife" . . . "infant child of" . . . "died in his 39th year." I found the shorter life spans and the high rate of infant mortality on the older markers. I identified the years when plagues scoured the island and the commonalty of the name Bligh—evidence that might attest to the promiscuity of the famous captain. Then I saw the one gravestone different from all the rest. There was nothing special about it. Some unfortunate sailor, probably from one of the whaling ships that worked the southern oceans, had died in the early eighteenth century. The significance of this one gravestone, however, was the timeworn etchings that showed the sailor's place of birth: Providence, Rhode Island, U.S.A.—the same town I was born in.

I felt a strange kinship for this unlucky sailor who once walked the same streets I did and died long ago and half a world away from family and friends. I rode back to the village preoccupied with dark thoughts and, unaccustomed to driving on the left side of the road, sideswiped an oncoming car whose driver had downed a few too many Australian beers. Luckily, I got off with a few minor cuts and bruises and a sprained ankle. I was still limping a few days later when I stopped in Honolulu on my way back to Florida and ran into Briggs in the lobby of the Surfrider Hotel.

"You probably angered the Trim God," Briggs said, when I told him about my run-in with the car. Having issued this profound statement in the way of explanation, he returned to his stack of postcards, leaving me to ponder the strength of my belief in omens and gods.

There was probably no other pilot to give the legend of the Trim God such weighty credence as Briggs. So diligent was he in his religious observance that when figuring his flight time for a crossing, he used an airspeed ten knots slower than he actually planned to fly. Since he would then arrive at his destination early, he was certain that guaranteed him a preferred relationship with the deity.

"Did you send your postcards?" he asked, while paging through his address book for more names.

"I wasn't planning on sending any," I said.

"You're not sending any?" he exclaimed, as he set his pen on the table and began lecturing. "I'm really disappointed in you. Sending postcards is an important part of ferrying airplanes. You wouldn't come out here without making airline or hotel reservations, would you?"

"No," I said, wondering where he was going with this. I didn't know if he was serious or joking. With Briggs it was often difficult to tell.

"Sending postcards is just as important. As a Whiskey Tango, you have an obligation to send postcards to your friends," he said, using the phonetic alphabet acronym for world traveler. "Friends who don't travel outside the country that much, and look forward to getting postcards so they can walk away from their mailbox with a smile on their face, clutching a postcard from South Africa or Hong Kong or some other exotic place."

"You're right," I said, humoring him. "How could I be so inconsiderate? I'm going to send some this afternoon." Then I noticed that one of the cards he was addressing had his own name and address on it. "I don't mean to question your logic, but how come you're sending a card to yourself?"

"I always send one to myself. Because when I get this card, then I know everybody else got the cards I sent them."

That sounded like something Briggs would do. I suspected it was really so he could walk away from his own mailbox with a smile on his face, clutching a postcard from South Africa or Hong Kong or some other exotic place. Our discussion of the duties and responsibilities of sending postcards ended when Kerby and Bennett walked into the lobby and joined us.

"Bennett? What're you doing here?" I said, surprised to see him.

"Right now, I'm wishing I was back in Lakeland," he sneered. Even when he was happy, Bennett looked angry. He could sneer better than anyone I knew.

"Bennett misses all the action, adventure, romance, and intrigue," said Kerby. "So Briggs and I talked him into taking this trip with us—just like old times."

"Yeah. The good old days," said Bennett. "Headwinds and broken airplanes."

"Bennett's not having much luck," Kerby said, consolingly.

"Don't tell me," Briggs said. "They didn't come in yet?"

"Floyd said maybe tomorrow," Kerby answered. "But I think he was just trying to cheer Bennett up."

"What's supposed to come in?" I asked.

"Just a few boost pumps, that's all," Briggs said. "Nothing to get excited about."

"Yeah, but they didn't break on your airplane," Bennett quickly added before telling me the story of how he got here in the first place. "We had these three Navajos come off the production line at Piper. We get all the paperwork, so I knew they were going to Sydney. Kerby told me he and Briggs were going to take two of them and starts telling me I should take the other one. I thought it might be fun to go on another trip. You guys had all these airplanes on the ramp, so when I asked Phil if I could ferry this Navajo he was only too happy to let me take it."

"Phil does get a little nervous if he looks out the window and sees too many airplanes," I said.

"So I started test flying the airplane I was going to take. And believe me, I put it through the ringer. I wrote up everything I could find wrong with it, and after the engineers fixed them, I test flew it again. It was probably test flown more than any airplane Piper built. But when I was through with it I was absolutely sure it was in top notch, A-1 condition."

"Aren't they all supposed to be like that?" Briggs asked.

"Let me finish," Bennett said. "So I started out on the trip and got about halfway to Honolulu when both boost pumps quit. At the same time! Then Kerby comes on and says 'Don't worry, you'll probably make it.' "

"Well, you still had the engine-driven pumps," Kerby said.

"But I didn't know if they were going to quit, too, and now I was starting to get worried."

"You worry too much," said Briggs. "You're here, aren't you?"

"Yeah, but for how long?" he said to Briggs before turning to me. "They told me it'd take a few days because they had to get the pumps shipped from Piper. That was four days ago. I've been going out to the airport every day to see if they've come in, but they keep telling me to check back tomorrow. At this rate, I may still be here next month."

"Oh, but we've been having a big time here on Wakiki," Kerby said. "The most fun you can have with your pants on. We even got Bennett started on pineapples."

"I meant to ask you guys about that," said Bennett. "When I woke up this morning, my gums were sore and its been getting worse all day. You think it could be from the pineapples?"

"Of course," said Briggs. "That's because of the acid in the pineapples. If you eat them all the time your teeth will fall out. That's why these natives don't have any teeth."

When I left the next morning, they were still arguing about boost pumps and pineapples, and Bennett was still complaining about his gums, to which Kerby kept telling him that was a small price to pay for all this action, adventure, romance, and intrigue. It was a saying he adopted as his very own and repeated whenever someone asked him what he was doing in such a dangerous business. In time the saying caught on with the rest of us, but no one said it with quite the same panache as Kerby.

To look at Kerby one would not suspect that beneath his calm exterior there beat the heart of a chivalrous knight ready to rescue the damsel in distress or do battle to slay the evil dragon. The trouble was, there were few damsels in need of saving and even fewer dragons. There were, however, a multitude of other problems that found their way to Kerby and myself in the coming months when I thought it was finally my turn to meet the Trim God.

12

An Albatross to the Far East

The Grumman Albatross was an amphibious aircraft built in the 1950s for the U.S. military and used for search-and-rescue missions. It had a cabin big enough to walk around in, required a two-man crew, and boasted a real lavatory—a marked improvement over our usual milk carton. Squatting on the ramp with its long wing span and pregnant underbelly, it bore more than a passing resemblance to the bird for which it was named. The Albatross had enough range to fly two to three times longer than most aircraft of its type, but in the age of jet engines it had become obsolete. Those that didn't end up at the air force graveyard in Arizona were brought back to the Grumman factory in Stuart, Florida, where they were overhauled and sold to third world air forces. That's where we entered the picture, because Grumman contracted with Globe Aero to ferry these aircraft to Indonesia.

Because no one at Globe Aero had flown anything as large as an Albatross, Waldman arranged for Grumman Aircraft to give each of his four senior pilots—Briggs, Kerby, Gray, and me—three takeoffs and landings in the airplane with a factory test pilot. Grumman, however, more intimately familiar with their product than we were, sent a mechanic with us in case we broke down en route. His name was Andy Mattenheimer, a burly ex-Brooklynite who had drifted through various occupations as a boxer and a bartender before spending twenty years with Grumman. For some reason he latched onto Gray and me and flew in our aircraft, probably because we appeared more likely to be in need of his agely wisdom than either Kerby or Briggs. He began our education with his views on natural selection.

"It's simple. Man survived because he stunk," he said. "Ya see, man isn't naturally clean like animals. If you put a man and a wild animal in the woods for a month, when you come back, the animal will smell okay because it's got the natural instinct to clean itself off. But without any place to take a bath or wash up, the man will come out smelling like hell."

"Okay, that's probably true," I said. "But what's that got to do with why he survived with all these carnivorous beasts running around?"

"Think about it. Man smelled so bad, none of these animals would eat him."

"What!" Gray exclaimed. "Andy, come on. What're you talking about?"

"No, listen. This is real easy to prove. All ya gotta do is go off in the woods for a month, and when you come back out, see if anybody wants to get close to you. That's what it was like in prehistoric days. We stunk so bad, these wild animals lost their appetite whenever they got downwind of us."

Mattenheimer's ideas about man's early history may have been suspect, but thankfully, his knowledge of the Albatross was more reliable. During the first leg, he was constantly checking fuel, hydraulic, and electrical systems, and keeping watch over the various engine gauges. I was handling the radio and navigation, while Gray was flying the airplane and getting a feel for the controls. Compared to the small aircraft we were used to flying, the Albatross was a behemoth. At thirty-five thousand pounds, it was heavy and sluggish, and unlike other aircraft that streaked or rocketed or soared through the sky, the Albatross lumbered.

It was stable, however, which we would be thankful for on the long flights over water. Gray appeared at ease at the controls by the time we landed at Kelly Air Force Base in Texas. Because both aircraft were painted in Indonesian air force colors, Grumman had made arrangements for us to refuel at U.S. military bases. Briggs and Kerby had landed a few minutes ahead of us and were refueling when we taxied next to them.

Taipei
Hong Kong
Macao
PACIFIC OCEAN
Bangkok
Manila
Saipan
Guam
Yap
Kota Kinabalu
Truk
Palau
Kuala Lumpur
Morotai
Singapore
Kuching
Balikpapan
Wewak
Jakarta
Surabaya
Malang

The Far East

"I think we should fill all the tanks now, so we can see how these airplanes are going to fly when they're heavy," said Briggs.

"It's awful hot," I reminded him. "The density altitude's probably a couple of thousand feet. We'll use up a lot of runway, and we're not going to climb very fast."

"It can't be worse than where we're going. Besides, I'd like to be over land when we fly them heavy the first time."

"We may as well top off, too, then," I said, and because it was Briggs, added, "Why don't you go first, though. That way when you crash on takeoff, I'll have some good stories to tell when we get back to Lakeland."

Briggs and Kerby used almost the full runway before becoming airborne. Though the Albatross looked as if it was barely moving, it did seem to be carrying the weight. Now it was our turn. I made a last minute scan of the engine instruments as I taxied onto the runway and pushed the throttles to full power.

The airplane sat for a second or two, vibrating and making a lot of noise, before it began to inch forward. It was a hot, muggy August afternoon. The air was so dense that the aircraft seemed to roll on forever as it gathered speed. Then the left wing began to drop, not much at first, but more and more as speed increased. As the wing got lower, the airplane began veering to the left. It took full right rudder just to stay lined up with the runway.

"What's the matter?" Gray asked.

"I don't know. It just doesn't want to stay straight."

The control tower was calling us, but at full power the engines were making so much noise that neither Gray or I could hear what they were saying. We accelerated past V2—the speed that the aircraft manufacturer says is the minimum for flight—and still the ground maintained its hold. The runway in front was decreasing quickly. I had the right rudder pedal all the way to the floor and the control wheel swung over to the right, but the airplane still kept pulling to the left. Something was wrong, but all the instruments looked normal and, although slowly, the aircraft was accelerating.

Finally, the overweight airplane reached a speed fast enough to support itself in flight and broke free of the runway. The left wing dropped even more, and I quickly counteracted with right aileron. The airplane gained speed slowly, allowing me to ease up on the right rudder. It didn't pull to the left, but we were still too slow and we weren't gaining altitude.

"Gear up!" I shouted. But before I finished the words, Gray had flipped the landing gear handle to the UP position. The sound of motors and jackscrews added to the din, as three thousand pounds of hydraulic pressure forced the heavy landing gear up into the wheel wells. Now streamlined, the aircraft accelerated quickly to a safe climb speed. As we passed through five hundred feet, I called for climb power. The control tower was still calling us when Gray reduced the throttles.

I still couldn't understand what they were saying, but I saw Gray look out his side window and then push the press-to-talk switch on the control wheel to excitedly shout, "We dropped an egg! We dropped a fucking egg!"

In his haste, Gray had pressed the wrong switch, transmitting not only to me, but to the control tower as well, because they quickly advised us to "Try to watch your language."

It took a few seconds before I realized the egg Gray was talking about was the three-hundred-gallon drop tank fastened to the wing. When I looked out the right side window, I saw that it was gone. I checked my side of the aircraft to see if the left drop tank was still there, but I knew it would be because the eighteen-hundred-pound imbalance between the wings is what caused the airplane to pull to the left during takeoff. Even now, I had to hold the control wheel to the right just to keep the wings level.

"Indonesian Air Force, what are your intentions?" the tower operator asked. "Do you want to return to Kelly?"

"Stand by while we try to figure this out," Gray said.

"I'm sure that drop tank shattered when it hit the runway," I said. "They're probably not going to be too happy with us if we land back there now."

"Yeah," Gray said, and gave an ironic snicker. "They've got all these students up flying, and we just closed their only runway. I'll just bet there's some second lieutenant down there with a ton of forms for us to fill out. I think we should get the hell out of here."

"One thing's for sure," said Mattenheimer, "we can't land with this much weight difference between the wings; we'd have to jettison the other drop tank. Either that or hold for three hours until we burned off fuel on the left side."

Deciding that it would be quicker to fly back to Florida for a new drop tank, we told the tower operator of our decision and asked him to advise Grumman that we were returning for maintenance. It was late when we landed at Stuart. We were tired from the long day of flying and wrestling with a lopsided airplane. We weren't prepared for the Grumman mechanic when he met us at the airplane door and told us, "Briggs is in Biggs."

"What?" I asked, not sure if the man was stuttering or if I was going deaf after listening to the two big nine-cylinder radial engines for the last several hours.

"Briggs is in Biggs. He lost an engine and landed at Biggs Army Air Field in El Paso."

While Gray, Mattenheimer, and I had our hands full with our airplane, Briggs and Kerby were fighting a losing battle with gravity. One engine, even at full power, wasn't enough to keep the heavy aircraft from dropping out of the sky. They had to get rid of weight, and there was only one way to do it. With one quick motion, Briggs pulled the disconnect lever and sent both drop tanks hurtling to the desert floor.

Thirty-six hundred pounds lighter and still losing altitude, the descent had slowed enough that they could make it to El Paso, if only the mountains weren't between them and the airport. As they were preparing for a crash landing in the desert, an army helicopter spotted their aircraft and led them through a pass in the mountains to the runway at Biggs just as the good engine overheated and quit.

It was only our first day, yet already both aircraft were broken down. Instead of being two thousand miles west of Lakeland, we were two hundred miles southeast. It was not a good omen.

Three days later, Gray, Mattenheimer, and I left the Grumman plant with a new drop tank fastened to our wing and two more in the cargo bay for Briggs and Kerby. Grumman had already sent a mechanic to El Paso, and by the time we arrived, he had found the ruptured oil seal that had caused the failure and rebuilt the engine. We spent a day waiting for Briggs and Kerby to recuperate from their two-day binge in Ciudad Juárez, Mexico, and then left for California.

The flight across the Rockies was notable only because when we landed at Alameda Naval Air Station in Oakland all four engines were still running, and we had the same number of drop tanks that we started with. So when the navy briefer told us we would have good weather to Honolulu, Briggs wanted to leave right away.

"The winds are supposed to be good tomorrow night," I said. "Maybe we should get some sleep first, so we can be wide awake for the crossing."

"That's what we've got two pilots for," Briggs countered. "You guys can take turns. That's what Kerby and I are going to do."

"But November's gone," said Gray. "I don't know about you guys, but this is my first time out here since the ship left. It's going to be a long flight without a beacon in the middle."

"What're you all worried about?" Kerby asked. "You fly to Pago Pago all the time with nothing in the middle—and that's a lot longer."

"With the problems we've been having, I'd feel safer if we crossed together," I said. "It's two against two, so that means Andy's the tie-breaker."

"It's up to you, Andy," Briggs said. "Do you want to stay here or would you rather, this time tomorrow, to be eating pineapples on Wakiki with all these cute little beach bunnies running around half-naked."

"And then we're gonna take you to the Monkey Bar at the Pearl City Tavern," added Kerby. "So you can watch the monkeys screwing the hell out of each other while you're having the best teriyaki steak on the island."

Two hours later we were flying over the Golden Gate Bridge and setting course for Honolulu. The Albatross was not as sluggish in the cold, night air, reaching a cruising altitude of eight thousand feet quicker than on previous legs. After assuring himself that the engines were operating as they should, Mattenheimer was fast asleep in the flight mechanic's seat. Waking periodically, he leaned forward to stare at the engine gauges and ask a question about pressure or temperature and, when satisfied his attention was not required, grunted and went back to sleep.

Gray and I, however, were too overwhelmed with the luxury of crossing the ocean in such a large aircraft to even think about sleep. No longer were we flying a small puddle jumper that barely got out of its own way nor were we sitting in a cramped cockpit alone for hours. We reveled in the position fortune had placed us in and gave our attention to the aircraft, except for the few times I went into the cargo bay to jump up and down and convince myself this was real.

As the novelty wore off, we took turns sleeping on the floor between the pilot and copilot seats. When daylight came and, with it, the airlines flying between Honolulu and the west coast, navigation was reduced to simply following the contrails to the island.

Honolulu Airport is a joint civil and military field with commercial aircraft on one side and military on the other, which was called Hickam Air Force Base. We parked on the military side and were refueled by the air force. Like our previous stops, no one asked us for identification, but simply handed us a clipboard and asked us to sign for the fuel. Our status as neither military nor civilian seemed to allow us a good deal of leeway, so Briggs decided to see just how much we could get away with.

"Are you air force?" the young airman at base operations asked, when Briggs told him we needed rooms at the BOQ.

"We're on a quasi-government mission and need quarters for three days," Briggs said, in his most authoritative airline-captain voice.

"And you are, sir?" the airman asked.

"Colonel Briggs."

The airman hesitated, appraising the rest of us suspiciously. In civilian clothes and with hair longer than the military allowed, we attracted more than a few curious glances from the air force pilots who probably didn't know what to make of us.

"Do you have your orders, sir?" the airman asked.

Briggs, enjoying the charade he had started, turned to Kerby and said, "Do you have the orders, major?"

"I've got them right here, colonel," Kerby said, and handed our Globe Aero ferry contracts to the airman. He studied them for a minute and, apparently satisfied, assigned us two suites at the Hickam BOQ.

As Kerby and Briggs promised, we showed Mattenheimer around our usual haunts in Honolulu. The morning of our last day found us at the pool bar of the Sheraton Wakiki, sitting under the giant Banyan tree and planning the rest of our trip.

"If nobody's got any particular desire to stay over on Wake Island, I think we ought to just stop there to refuel and fly to Guam," I suggested.

"But that'll be twenty hours of flying," complained Kerby.

"I guess we could spend the night on Wake," Briggs said, in his usual sarcastic tone. "I read somewhere that they've got a Quonset hut built during World War Two that they use for a hotel. You're a whiskey tango, Kerby; I'm sure you'll be quite comfortable there."

"Well . . . uh," Kerby stuttered.

Ever the diplomat, Briggs suggested we put it to a vote.

"You guys decide. Either way's okay with me," said Gray.

"I think we should keep going," Mattenheimer said. "Seems like we've been on this trip forever and we've got what, another seven or eight thousand miles to fly. I don't see how you guys ever get to where you're going if you have to vote every time you do something."

"Democracy ain't easy," I said. Although either aircraft could fly whenever the crew wanted, because we had decided earlier to fly the trip together, we now had to deal with the democratic process. With two or three people it was difficult; with five it was impossible without a lengthy discussion of the merits of everyone's thoughts, which did not end until we left that evening for Wake.

Except for the steady drone of the two fourteen-hundred-horsepower radial engines, the night crossing was quiet. It was my turn at the controls, which left Gray and Mattenheimer free to while away the hours scanning the channels on the high frequency radio. Both Albatrosses had new, solid-state radios with a fixed antenna that we didn't have to crank in and out every time we changed frequency. They were a vast improvement over the ancient, vacuum tube Sunairs we usually flew with, and Gray was apparently enjoying this new toy.

Several channels of music came and went before Gray stopped on one frequency that, because of the crisp, British accent of the announcer, I assumed was the BBC. After a few minutes of listening to the latest production quotas for tractors and wheat tonnage, I concluded it couldn't possibly be the BBC and asked Gray what we were listening to.

Gray was an enigma—blond, Irish, and so apple-pie American that when he was a kid he was a member of Howdy Doody's Peanut Gallery. He had told me that he went to Moscow when he was a teenager but wouldn't say much about it. His apartment was decorated with posters of Lenin and the Red Army, and the Royal Canadian Mounted Police in Gander once questioned him because they thought he might be working for the Irish Republican Army. Of course Gray was neither a terrorist nor a communist, but he liked to keep people guessing. The pale, red glow from the instrument panel bathed the cockpit in a sinister, almost surreal light, as he sat down, snickered, and said, "Radio Moscow."

From the air, Wake Island looks like a snake coiled up and resting. Had the sun been directly in front of us, we might have flown past the tiny atoll without ever seeing it. The airport wasn't much—an old Quonset hut for a passenger terminal and one hard-surface runway long enough to handle the Boeing 747s that flew overhead but never landed. In the early days of transpacific aviation, Wake Island was a refueling stop for the Pan American flying boats and later the propeller-driven Constellations and Stratocruisers flying between Hawaii and the Far East. Then came the jet age and the Boeing 707s and Douglas DC-8s with intercontinental range, and Wake Island was abandoned by the airlines and left to the almost exclusive purview of the military. It was an outdated relic of a shrinking world. Its place was with the legends of aviation lore. One legend in particular, remembered fondly by the old timers, is that of the shark known as Mag Check Charlie.

In the days before the jet engine, airliners were powered by big, reciprocal engines that burned oil, backfired, smoked, belched, and probably acted like the senile old uncle you try to hide when guests come to visit. The engines had two spark plugs per cylinder connected to two independent magnetos, one for the right bank of plugs and one for the left bank. Before takeoff pilots checked each magneto by revving the engine to a high rpm and momentarily switching from the right magneto to the left magneto before returning to the BOTH position. If the engine ran smoothly on each independent magneto, flight could be commenced with no undue concern. But if engine power started breaking down, it indicated a problem with the ignition system that could result in a power failure on takeoff. On Wake Island, with the runway running along the beach, if an engine quit during takeoff there was only one place the airplane ended up—in the water.

As legend has it, Mag Check Charlie swam close to the run-up area whenever he heard an airplane taxiing out and then waited for the pilot to perform the magneto check. If the shark heard a large drop in rpm, it swam to the opposite end of the runway to be in place for the inevitable feast when the airplane crashed.

If there was any truth to this legend, we weren't able to find out, because both Albatrosses' magnetos passed this simple test and we were soon in the air again. It was an all-day flight to Guam, hot and sticky with the equatorial sun beating down through the large cockpit windows. By the time we landed, we had been up almost twenty-four hours, had flown more than thirty-four hundred miles, and were not exactly at our best mentally or physically. Even with two pilots in each aircraft to share the flying, the back-to-back legs had been hard work. We were tired, grouchy, and smelly, when we walked into the Officers' Club and the waitress told us we had to order quickly because the kitchen was about to close.

"This is ragtime," complained Kerby, which got the rest of us griping and cursing the uncompromising marriage of slow airplanes, vast distances, and restaurants that closed early. Mattenheimer, unusually quiet except for a few growls, was making ugly faces in the wall mirror at the end of our booth.

"Andy, uh . . . what're you doing?" Gray asked.

"I'm getting rid of all this tension," he said. "This is what you guys need to do. You're all too uptight. You gotta look in the mirror and make ugly faces. I do this all the time when I have a hard day."

At this point, I was punchy enough to try anything. Looking into the mirror, hesitantly at first, I gritted my teeth and sneered at my reflection. "Like this?" I asked.

"Now ya gotta growl," Mattenheimer instructed.

"Grrr!" I growled, and was quickly joined by Gray, Briggs, and Kerby, all sneering and growling like a bunch of Neanderthals after a long hunt. Although we probably all looked like madmen when the waitress returned with our food, it did seem to be having the calming effect Mattenheimer claimed it would because we were no longer complaining.

The following evening we took off for Balikpapan, an Indonesian air force base on the southeast coast of Borneo. After five hours, contrary to the satellite picture we had seen on Guam, we were still in the clear. The radio was quiet, so I figured Briggs and Kerby, a half-mile to my left, were taking turns sleeping as we were doing. Gray was stretched out on the floor between the pilot and copilot seats, and Mattenheimer was slouched in his seat with his head down.

It was that time of the crossing when one is lost in the solitude of his innermost thoughts. The trip was almost over and I felt the old, familiar feelings of anticlimax. After a quick refueling stop in Balikpapan and then a short hop over to Malang, we would be through—ten more hours of flying, twelve at the most. I was content with the boredom and let Gray sleep past the time he had asked me to wake him when the left engine began losing power.

"What's happening?" Mattenheimer grumbled, leaning forward, his eyes quickly scanning the gauges for some clue.

"I don't know. She just started losing power and oil pressure. I increased the mixture, but the rpm's still dropping. I'm going to shut it down while we've still got oil pressure."

As I moved the throttle forward for the right engine, Mattenheimer pushed in the feathering button for the left engine, which rotated the propeller blades into the slipstream. Seconds later, the big barn door of a propeller slowed to a stop. The airplane yawed hard to the left and airspeed dropped drastically. Still heavy, and with only half the horsepower, we began losing altitude. I pulled back on the control wheel to stop the descent, but the airplane started shaking. I could have either speed or altitude, but not both. The only thing I could do was tread a fine line between stable flight and a stall, and settle for the slowest possible descent.

"What in the world are you all doing down there?" Kerby's disembodied voice started through my headset.

"We had some problems with the left engine, so I thought we better cage it for a while," I said, trying to sound less worried than I was.

"What can we do to help?"

"I don't know yet, Kerby. We're kind of busy right now. I'll get back to you as soon as we get things straightened out."

"Okay, but don't worry about navigating. We're slowing down, so just keep us in sight."

All the commotion must have awakened Gray, because he was climbing into his seat and rubbing the sleep from his eyes. It was not the ideal way to wake up: red lights flashing, wings aslant, and the windshield filled with a black, starless void, part of which was ocean.

"The drop tanks!" he said. "They're still full aren't they? Maybe we oughtta get rid of them."

"Let's wait a while. I'd like to keep the airplane in one piece if we can. Besides, at the rate we're losing altitude, we've probably got thirty or forty minutes before we have to do something. Maybe we'll burn off enough fuel so we can maintain altitude."

"Let's just not wait too long," he urged.

As the Albatross lumbered through five thousand feet, the descent had slowed down, but we were still losing altitude and would eventually end up in the water if we didn't do something.

"Do you still have us in sight?" Briggs asked. "We can't go any slower. Can you follow us okay?"

"We can't right now," Gray told him. "You're pulling away."

"Okay. I'll make a three-sixty so we don't get too far ahead. Kerby put out a Mayday for you with Clark Radio."

"Are they going to send anybody out?" I asked, somewhat hopefully.

"I don't think so," said Kerby. "The guy asked me what was I telling him for. He said we were a search-and-rescue airplane and were already here, and there wasn't anything he could do that we couldn't."

"Well, are you guys ready for some more water-landing practice?" I joked . . . sort of.

"Oh, me! Now don't go telling me that," said Kerby. "As dark as it is, we'll probably kill ourselves trying to put this thing down on the water."

"If we have to ditch, I'm just glad we're in a seaplane," said Gray.

"I'll tell ya, we just might have to do that," said Mattenheimer. "Look at the cylinder-head temp for the right engine. It's too close to redline. We gotta get the temperature down or it's gonna overheat."

The airplane stopped descending at eleven hundred feet. We were just above a stall at ninety-five knots, but at least we were maintaining altitude. Now if only the right engine didn't overheat we might be able to avoid having to ditch.

"We're going to melt the engine if we keep this up," Gray said.

"It feels like we're light enough now. I think we can ease the power back a little," I said, while moving the throttle a fraction of an inch. Manifold pressure reduced slightly, and a few minutes later the cylinder-head temperature dropped into the green arc—within limits but still dangerously high.

"We'll never make Balikpapan at this rate," said Gray. "We have to find some place closer."

As usual, Kerby was one step ahead of us. "The closest airport I can see on the chart is Pitu on the island of Morotai," he said. "It's about five hundred miles away, and it doesn't have a beacon. But it's part of Indonesia, so at least we can get the airplane to the right country. It's up to you guys, though."

"Doesn't look like we have much choice," I said. "Can you guys lead us there? With this engine caged, I don't know how accurate our compass is."

"Yeah, just follow us. I got it all plotted. I figure we should get there just before sunrise."

Finding an island five hundred miles away was difficult even if we knew where we were at this exact moment. At best, all we had was an approximation of our position. We would have to go from one estimated position to another estimated position and then try to find an island, visually and in the dark. It required considerable faith in the forecast winds, in Kerby's skill at dead reckoning, and in our remaining engine.

As we burned fuel and the airplane became lighter, we were able to keep the engine from overheating by reducing power. But this was creating another problem. Because the right engine was using fuel from the right wing tank, that side of the airplane was getting lighter while the left stayed just as heavy. And the longer we flew the worse it got. We had turned the autopilot off because we had to fly with the left wing five degrees higher than the right to compensate for the drag from the feathered propeller. But with no horizon for reference, Gray and I could take only thirty minutes each at the controls before we started wandering.

Three hours later the fuel imbalance between wings was almost a thousand pounds, and it was becoming more difficult to keep the airplane going in a straight line.

"We gotta switch tanks," Mattenheimer urged.

"But what if the engine quit because we got bad fuel at Guam? If we switch tanks, we'll be sending that same fuel to the right engine."

"If we keep flying like this it won't matter because we'll overstress the wings," Mattenheimer warned. "But I don't think it was bad fuel."

"What makes you say that?" asked Gray.

"It just didn't seem like it," Mattenheimer said, as he grabbed a flashlight from his flight bag. "I'm going back to see if I can tell anything from the side window."

He came back into the cockpit five minutes later. "I couldn't see much because of the dark, but it looks like there's oil all over the cowling. It has to be an internal failure."

"We have to do it," Gray said.

"I know, but we're gonna have to be quick. Because if it starts to quit, we won't have much time. Everybody ready?"

Gray's hands were on the mixture and prop levers, Mattenheimer's finger was poised a hairsbreadth from the feathering button, I had one hand on the control wheel, and with the other, I reached down to the fuel control panel and switched tanks.

The next five minutes dragged by. Another five minutes passed before I relaxed my grip on the control wheel and Gray and Mattenheimer abandoned their hovering over the engine controls. We were still in the air, and the engine was still running. The three of us looked at each other, the relief evident in our faces, and started laughing over our luck.

The wings were just about balanced when we neared our ETA for the island. Sunlight was beginning to lighten the eastern horizon, but in front and below, it was still so dark we could already be over the island and not know it. Although unlikely, it was still a possibility; so we began searching for anything that suggested a shoreline, a road, or lights.

"I think I see something!" I said, barely able to hide my excitement.

"Where?" Gray asked, as he stared at the blackness.

"Right below us. It might just be moonlight reflecting off the whitecaps, but I think we should take another look." We were still at eleven hundred feet, an altitude that, according to Kerby, was a thousand feet below the hilltops. I banked the airplane into a wide circle. Gray and Mattenheimer were glued to the windows trying to distinguish shapes from the black depths of the ocean.

We spotted the shoreline on the third circle. Not only had Kerby estimated our time almost to the minute, but the course he plotted was so exact, it took us directly to the center of the island. I was amazed at the accuracy of his navigation, until I was struck with a sobering thought. If we had gotten here five minutes sooner it might not have been light enough to see the shapes on the surface, and we would have kept right on going into the side of a hill.

After praising Kerby's wizardry at dead reckoning, we asked where the airport was located on the island. Gray and I had only instrument charts void of any topographic features or small airports. Kerby, as usual, had a flight bag full of almost any chart one might possibly need.

"It's on the north end of the island," Kerby advised us. "But it doesn't look like it's paved—probably not more than a runway hacked out of the jungle. The chart I got is about two years old, so there's no telling what condition the runway is in. We're gonna circle overhead, so let us know when you all see something."

It was daylight by the time we reached the northern end of the island. The sky was clear, and there were almost no clouds blocking our view as we looked for the airport. Nourished by the rains and rich volcanic ash, thick jungle covered most of the island.

"I don't even see a clearing," Mattenheimer said, as I began realizing our worst fear, that the runway had been reclaimed by the jungle and was now overgrown with trees.

"I'll bet it's over there," said Gray, pointing to a small patch of low-hanging stratus. There were no other clouds in sight except for that one small spot on the extreme northern section of the island.

"That'd be typical," I said, while turning the airplane in the direction Gray had indicated. Sure enough, as we descended through the thin cloud, the runway came into view. We made a low pass to check for obstacles, made another circuit, and landed.

The airport looked deserted at first, but as soon as we shut down the engine, people started coming out of the jungle from every direction. A jeep approached from the opposite end of the runway with three men waving what looked like machetes.

"This doesn't look good. We better tell Briggs and Kerby," I said. Gray was already on the radio, advising them not to land because the situation did not look promising.

A crowd was standing around the airplane when the jeep braked to a stop and the three men got out. One of them wore a military uniform; he stood by the door and just waited, apparently for us to make the next move.

"We might as well get it over with," I said. So, not knowing what to expect, we opened the rear door of the aircraft and walked outside.

Indonesians are small, slim people. Gray and I easily stood three or four inches over the tallest of them. Mattenheimer, big, burly, with a great girth of a stomach, towered over them, especially the children who followed him around as he stormed out of the airplane with a stepladder in one hand and a tool bag in the other. The small Indonesian children were fascinated with this giant who dropped out of the sky and was now peeling open the engine cowling. They jumped at his every request for tools, transcending the boundaries of language, as they handed him a crescent wrench or a pair of pliers.

"I am commander of airport. Sergeant, Indonesia army," said the guy in the uniform, while pointing to his stripes. "We not see airplanes here many years."

That explained the crowd, whose curiosity seemed to go no further than gawking in fascination or touching this glistening bird that had descended upon their remote world. Satisfied the situation was not as dangerous as we first suspected, I called Briggs and Kerby to let them know we didn't get shot at, so it was probably safe for them to land. By the time I went back outside, another group of natives different from the rest had arrived. They were dressed in fancier clothes, for one thing. But what really set them apart was the white paste the women had on their faces. We learned later that they were part of a wedding, which was interrupted because of the excitement going on at the airfield.

"Well, this airplane's not going anywhere for a while," said Mattenheimer, after closing up the engine cowling with the aid of his loyal disciples.

"What's wrong with the engine?" I asked. Mattenheimer said something that I couldn't hear because Briggs chose that instant to make his entrance. He came in low, just clearing our airplane, landed, and turned around at the end of the runway. The crowd of curious onlookers shifted their attention from us to the noisy airplane that was pulling up next to ours. The white-faced girls and their group pointed to the airplane and chattered among themselves. Everyone was looking at the two pilots in the cockpit, waiting for something to happen, when Briggs opened the side window, stuck his head out, and announced, "You're probably all wondering why I called this meeting."

The Indonesian archipelago is a wide smile of islands scattered throughout the Pacific, south of India to Australia. Comprised of more than seventeen thousand bits of land ranging in size from Sumatra, the sixth largest island in the world, to the tiny Moluccas, the entire archipelago stretches a distance of thirteen hundred miles, three time zones, and two monsoons. If there was a word to describe this nation threaded together by straits and channels, it would be *diversity.* There are more than four hundred volcanoes, fifteen hundred species of birds, and nearly thirty thousand plants. In the midst of this varied environment, the cultures of Indonesia are noticeably different from island to island. There are over three hundred languages spoken by the numerous ethnic groups that inhabit the islands.

It was no wonder that the commander of Abdul Rachman Seleh Air Base on the southeast part of Java was disappointed when we told him that we had to leave one of the Albatrosses on Morotai.

"This is not good," he said. "The army unit on Morotai backed the wrong side during the last coup. The air force does not know how loyal they are. I must call headquarters in Jakarta."

But every time he tried, all he got was a scratchy voice that kept saying "Hello, hello" and then hung up.

Kerby shook his head in amazement. "What happens when a war breaks out?" he asked. "How do you all get through?"

"We don't," said the commander. He tried a few more times before giving up in frustration. "Bad lines. We cannot get through now. You must clear this up in Jakarta."

Telephones were not the only thing that needed modernizing in Indonesia; the other was its military, hence our small role in the name of progress. There were a few prop-driven transport planes and small utility aircraft parked on the flight line. What most caught our attention were the five Russian MiGs left over from when Suharto seized control of the army in the mid-1960s and kicked out the communists. No longer flyable because of a lack of spare parts, the commander let us take a closer look at them, snap pictures, and climb all over them like kids let loose in a toy store.

If ever there were a contrast between modernism and tradition, Indonesia was the ultimate paradox. Each stop was like a march through time: the old-wedding traditions of Morotai; the airbase trying to modernize its fleet; and Surabaya, the industrial center and port with its cockfights and martial-arts movies imported from Hong Kong. The contrast continued the next day with our short flight on Garuda, the national airline named after the mythical half-man, half-bird of Indonesian folklore. And there was Jakarta, with its shopping malls, traffic, and growing capitalism.

After checking into the Intercontinental Hotel, Gray called Waldman to let him know what had happened and I got hold of Grumman's agent, an American, who sounded not at all upset over the situation. The following morning, the agent and I—as the senior pilot I was sort of the unofficial flight leader—met with the air force general staff who weren't any happier than the officer in Malang at getting only one airplane instead of the two they were expecting. Though disappointed, they agreed that leaving the airplane was the only thing we could've done and that they would get it fixed and fly it to the base themselves. With the meeting concluded, I returned to the hotel to join my comrades for the taxi ride to the airport and the flight to Hong Kong.

Before leaving Lakeland, we had made reservations on Japan Air Lines to fly back to the United States by way of Hong Kong. Because we would arrive in Hong Kong too late to catch a Japan Air Lines flight to the United States, we qualified for a free overnight. When we arrived in Hong Kong and tried to check in at the Hyatt Regency, however, the desk clerk wouldn't accept the airline vouchers.

"Whad'ya mean you won't accept them?" Kerby demanded.

"We are very busy tonight," the clerk said. "The airline rooms are all sold out."

"I don't care how full you all are. These are legitimate vouchers," argued Kerby.

"You must take that up with Japan Air Lines. We have rooms, but we cannot take these vouchers. You must pay for them."

"But we already paid. That's why we have the vouchers," Kerby argued.

The lobby was busy with late-arriving guests, several standing in line behind us, when I remembered Kuney's advice in dealing with airline ticket agents. Threatening didn't often work with the airlines, but it might with a hotel clerk.

"Look, we arranged all this weeks ago. We just came in on the JAL flight, we've been flying all day, and we're tired. We're leaving tomorrow and are supposed to have rooms at this hotel. Here, these are our tickets." For some reason I was holding everyone's ticket, which I laid out on the counter as proof of our claim. "Now are you gonna give us the rooms or what?"

Kerby's hands were in his pockets jiggling change, his neck was twitching, and he kept saying, "This is ragtime!"

Mattenheimer too was becoming frustrated, because he finally blurted out, "Don't argue with the son of a bitch. Just hit 'em!"

Either the clerk wanted to avoid a scene in front of the hotel's guests or he was afraid this burly American might actually carry out his threat, because he laid out five registration cards, scooped up the vouchers, and handed us five room keys.

I awoke staring at the ceiling. My eyes darted from one corner to the next until circumnavigation brought me back to the starting point. Something was troubling me, but I didn't know what. I got out of bed and went straight to the dresser, passing over the loose change and keys I had emptied from my pockets the night before, and picked up the envelope containing our five airline tickets. Tearing open the envelope, I pulled out the one and only ticket and read the name PAUL BRIGGS.

I searched the pockets of the clothes I had worn yesterday, checked my flight bag and suitcase, and looked under the mattress where my wallet was hidden. The chain was still on the door, so no one had gotten inside the room and stolen the other tickets. But where were they? I tried to think back to when I last saw them. It had to be when we checked into the hotel, because I remember laying them down on the reception desk. A call to the desk clerk, however, confirmed my dread, when he told me that no one had found any airline tickets.

Briggs, Kerby, and Gray took it pretty well when I broke the news to them over breakfast. Even Mattenheimer didn't seem too upset, although he rolled his eyes toward the ceiling and shook his head in disappointment as his plans for a quick return trip home evaporated. "I knew I shoulda never let you guys handle my airline ticket," he said.

"Andy, you don't have to get back to Pinellis Park that bad," I said. "Your wife's probably glad to get rid of you for a few weeks."

"Trips like this don't come along that often, Andy," Briggs began his sermon, "so we have to cram all the adventure we can into this one. You can't come all the way out here without seeing Hong Kong. Not if you want to be a real whiskey tango."

With Briggs's promise of adventure, Gray and I urging him that we had to stick together, and Kerby assuring him that we were all going to have a big time, Mattenheimer gave in. "I just wanna know one thing," he said. "When are we gonna leave?"

We didn't find out the answer to that question until Kerby called Japan Air Lines and was told it would take three days to trace the tickets and cancel them before new ones could be issued. While none of us would admit it, especially to Mattenheimer, we were glad for the delay because it gave us an excuse to take in the city.

"Look on the bright side, Andy," Gray said.

"What bright side? There isn't any bright side."

"Of course there is," I said. "This is an omen."

Briggs, holding his ticket in front of him for Mattenheimer to see and envy, added, "Anyone want to buy an airline ticket?"

"You're not gonna go back and leave us here?" Kerby challenged.

"And miss out on all this action, adventure, romance, and intrigue? You gotta be kidding."

Among the nations of the world, Hong Kong was unique. It was a European colony when there were few European colonies remaining. It was a citadel of capitalism sitting on the doorstep of the largest communist country on earth. Comprising less than five hundred square miles, Hong Kong was densely packed and crowded with high-rise apartments sprouting from the jagged mountains that surround the colony. The place reeked of unimaginable wealth and people—an overwhelming crush of people, most of whom seemed to be in a race with time or in a desperate rush for money.

Seeking refuge from the crowds, we found our way to the Peninsular, the colony's landmark hotel. After treating ourselves to that most sacred of English colonial tradition—high tea in the opulent marble and gilded lobby—we continued our exploration of the city, eventually arriving at the harbor.

Hong Kong owes its existence as a trading center to its deep harbor, which can easily handle vessels of every description from sampans and junks to ocean liners and supertankers. We took the ferryboat to Hong Kong Island and the sprawling Chinese marketplace in the old section of the city. On either side of the endless twisting streets were countless stands and tiny shops, each offering a maddening display of clothes and cheap jewelry, intricately carved wooden figurines, and hand-painted petits. We wandered past bustling markets and outdoor stalls with Chinese diners bent over bowls of steaming rice and tourists bargaining over the price of some worthless trinket hidden among the piles of merchandise.

"There's gotta be an opium den around here somewhere," Gray said.

Briggs remarked that he didn't know about an opium den, "but I'll bet we could find our airline tickets here."

The following morning we boarded the hydrofoil, a sleek new addition to the Star Ferry's fleet of aging ferry boats, and made the forty mile journey down the South China Sea to Macao in just over an hour. Until today, neither Gray, Mattenheimer, or I had ever heard of Macao, but Briggs and Kerby seemed to know all about the small island and were surprised at our ignorance.

"You gotta have heard about Macao," said Kerby. "The place is famous."

"Clint Eastwood made a movie there," offered Briggs. "I don't remember which one, but he wasted an army of Chinese drug dealers and thugs."

Suspecting that Briggs's memory might be faulted due to his preference for movies starring the famous actor, I suggested, "Yeah. That must've been *A Fistful of Macaos*."

I didn't know if Clint Eastwood ever made it to Macao, but the Chinese did. Macao is the oldest surviving European settlement in Asia. Its history as a trading center between Europe and the Far East dates back to the sixteenth century. Now its major industry is casino gambling, and it exists largely to satisfy the Chinese passion for games of chance. Every casino we went into was packed with Chinese standing several deep at the blackjack tables and slot machines, waving their money and clamoring for a chance to beat the odds.

Quickly tiring of the crowds, we hired two rickshaws to take us around the city. While there may have been a distinct absence of anyone that even remotely looked Portuguese, there was no mistaking the historical marriage of the two cultures. We rode past Roman Catholic churches standing next to Chinese temples and down cobblestone streets with Portuguese names lined with the shops of Chinese merchants. In every store, restaurant, hotel, and casino, even on the front page of the local Chinese-language newspaper, were pictures of Mao Tse-tung. We learned later that Chairman Mao had just died.

Preoccupied with the centuries-old sense of adventure and intrigue of this tiny enclave, we completely forgot about the time and got back to the port too late to catch the last hydrofoil. All that was left was one of the older and slower ferryboats that chugged through the water at a miserly eight knots and took five boring hours. We found a few empty seats on the outside deck in the aft section of the boat and settled in for the long ride. The wake we cut in the water was little more than a burble compared to the ferocious churning the hydrofoil had produced this morning.

After watching the unchanging shoreline for two hours, Gray had apparently thought about the situation enough to remark with knowing confidence, "I'll bet this is the slow boat to China everybody's always talking about."

The lights of Hong Kong were just coming into view when Briggs and I sneaked up onto the roof of the ferry. In true whiskey tango fashion, Briggs laid out a white linen tablecloth and produced two wineglasses along with a bottle of Portuguese wine he purchased in the port's duty-free shop. The river was void of traffic, save for one small sampan silently passing in the opposite direction with its sole Asian occupant. He reminded me of us in our tiny Pipers and Cessnas happening upon a large airliner over the vast expanse of ocean.

That, Briggs and I speculated, was mostly in the past. The gods had rewarded our patience by allowing us the chance to fly real airplanes. This was just the beginning. Grumman had two more Albatrosses for us to ferry and after that, who knows. Having proven ourselves capable of handling large aircraft, more were sure to follow.

13

Escape

I ferried two more Albatrosses to Indonesia over the next four months and still the air force hadn't retrieved the one we left on Morotai. Those subsequent trips—one with Gray, the other with Kerby, and both with Mattenheimer—went more smoothly than that trouble-laden first trip, and all too soon I was crossing the ocean again in small airplanes. What we had hoped was the first in a flood of contracts to ferry large aircraft turned out to be merely a trickle. Globe Aero delivered a few Metroliners and Shorts, which Briggs ferried because he had flown both aircraft during his commuter airline days.

Although the Albatross trips were few, they had a major impact on us in ways greater than any prestige we might have felt from piloting large aircraft. Following the minivacation we took in Hong Kong, on the next two trips we stayed over in Taiwan and Bali respectively, all at the airline's expense.

With the price of airline tickets rising an average of ten percent each year, we were always looking for ways to reduce the cost of traveling, legally if possible, and when that didn't work, we came up with some scam that bordered on the felonious. We made fake airline ID cards and rode in the jump seat for free, and we bought tickets originating in one country but used them in another. The times we most often used this strategy was when we were delivering an airplane to Frankfurt, where, because of the strong deutsche mark, airline fares were expensive. Instead of buying a ticket originating in Frankfurt and paying in converted marks, we bought one from Rome to the United States with a stopover in Frankfurt and paid in converted lira, a much weaker currency. Only we wouldn't go to Rome; we simply threw away the Rome-Frankfurt portion of the ticket, checked in at the airport in Frankfurt, and pretended we had just flown in from Rome.

Manipulating currency exchange rates was pretty sophisticated stuff, yet it was child's play compared to the ruse perpetrated by one enterprising pilot who formed an airline that existed solely on paper. So thorough was his deception that a flight schedule for the non-existent airline was listed in the *Official Airline Guide*. He then entered into an interline agreement with British Airways, which enabled him to fly for one-fourth the price of a regular ticket. His scam eventually ended when an alert station manager became suspicious of this airline that neither he nor anyone else in his office had ever heard of and began asking too many questions.

Airline ticketing is a complex system of codes and acronyms that rivals anything conjured up by the Pentagon: open-jaw, FCU, mileage break, city-pairs, and legal stopovers, to name just a few. While many of us knew what these terms meant, we had no idea how they related to actually constructing a ticket, let alone how we might benefit from this knowledge. Most of us were satisfied to fly on whichever airline could get us where we wanted to go the quickest and the cheapest. If we gave any thought at all to our travel plans, it was to try not to fly on the same airline more than once a month so we wouldn't have to watch the same movie.

Kerby, however, changed all that. No doubt influenced by the Albatross trips, he took it upon himself to learn not only the airlines' complicated ticketing process, but also how we could use that knowledge to our advantage. The pilots' room at Globe Aero's hangar—once a sanctum of charts, flight manuals, and a host of global-operational publications—soon began to look like a travel agency, as airline guides, hotel directories, travel brochures, and various magazine articles on remote getaways claimed more and more space.

Flight planning took on a new dimension. Choosing the return airline flight was now as important as weather and route selection. Before checking on weather conditions for the crossing and before even calling our travel agent, we first consulted with Kerby to find out what new and interesting excursion he had just discovered. It wasn't long before we were flying from country to country and scheduling stopovers so that we would miss connections and the airline would have to put us up for the night.

One of the more interesting of these excursions was the SAS flight from Bangkok to New York with an overnight stop in Copenhagen. Depending on the day of the week, the aircraft made a refueling stop in Tashkent in the Uzbek Republic, USSR. The aircraft was on the ground only long enough to refuel, but the passengers had to deplane and were herded like school children to a large, graying, cinder-block building that housed the duty-free shop.

Everything about the place looked old, run-down, secretive, and military; even the matronly *babushka* guides who kept us in line, turning often to look back at her charges to make sure we were still grouped into one herd and following her. Occasionally she waved her arms while shouting frantically in Russian if anyone strayed too close to the heavy iron fence that separated the airport from the alien—and for us at least—forbidden Soviet Union.

Forty-five minutes later everyone is herded back outside, past the Illyushins and the Antonovs painted in Aeroflot livery, to our refueled and waiting DC-8. Again the frantic arm waving and the unmistakable sound of a rifle being cocked if anyone dared bring out a camera.

SAS operated two different aircraft on this flight: a McDonald Douglas stretch DC-8 one day and a wide-bodied DC-10 the next. Because neither aircraft had the range to fly nonstop from Bangkok to Copenhagen, they stopped for fuel in either Tashkent or Tehran, Iran. Different aircraft and different fuel stops, the logistics of which was determined, not in the operations center of a cost-conscious airline, but in the cold war mentality of the Kremlin. The DC-8, which was the same size as Aeroflot's largest airliner, was allowed to land in Tashkent, but the DC-10 had to fly the much longer route through Tehran because the Russians didn't have a comparable wide-bodied aircraft and wouldn't allow it to use their airports.

Of all the airlines we flew, the one favored by most of us was Pan Am, which we dubbed "Pandemonium" because it was already becoming weighted down with the bureaucratic inefficiency of its globe-girdling operations. It was that same route system, however, with more flights to more countries than just about any other airline in the world that we could always count on to get us out of some place in a hurry. It also had the best airport clubs. We spent so much time at airports that many of us eventually joined at least one of the airline clubs. The most popular was Pan Am's, and the clubroom we frequented most was the Clipper Club lounge at London's Heathrow Airport.

Heathrow was like a small city with its own police and fire stations, chapel, hospital, shopping center, restaurants, hotels, and post office. It even had its own resident ghost, something that could only happen in a country whose news media aired reports on pasta crop failures and extra terrestrials drawing circles in crop fields. So why not a ghost? While investigators of the paranormal, along with a gullible populace, might attach some credence to this latest rumor, I suspected it was more likely some unlucky passenger who got lost wandering the maze of passageways and corridors until he was swallowed up into the bowels of the airport and never seen again.

I heard about this alleged ghost when I was stranded at Heathrow because of the transportation strike in Paris. This was nothing out of the ordinary; the socialist countries in Europe were often on strike for some reason or other. Mail, transportation, garbage—nothing was sacred. If it wasn't the Italians, it was the Brits or, like now, the French, whose civil-service strike put much of European air traffic into turmoil.

I was killing time in the lounge, waiting for my flight to New York to be called, when in walked Gray, looking tired from crossing too many time zones in too few days. He had just delivered an airplane to Germany and didn't want Waldman to know he was on his way back to Lakeland.

"Phil's got an airplane for Pat White, and I know he wants me to take it," said Gray. "I don't want to go out to the Pacific. Not for a couple of days, anyway. So whatever you do, don't tell him you saw me."

"Why don't you just take the trip and make a vacation out of it?" I suggested. "Get Kerby to come up with some weird place to go to after you get to Singapore. You could probably stay over in Russia."

"Except the airplane's not going to Singapore. It's supposed to go to the Philippines."

"The Philippines can't be too bad. Kerby just showed me an article about it in one of his travel magazines. He says it's a great place."

"Kerby thinks any place he's never been to is great," Gray said. "I've got a funny feeling about this trip. First the airplane was held up because the banks were playing with interest rates, and now Kerby tells me Phil's saving it for me."

Besides his expertise with airline fares, Kerby was also our spy. With his network of contacts, he often knew well in advance what airplane was going where and which pilot would be taking it. I had already been contracted to ferry an airplane to South Africa, so when I got back to Lakeland, I figured it was safe to go into the office. The first thing Waldman asked was if I knew where Gray was.

"I don't know," I lied. "I thought he was on a trip."

"He should be back from Germany by now," Waldman said. "I want him to take this Philippines airplane."

"Maybe he stopped in Gloucester on his way back."

"I already called his mother. She thought he was still in Europe. I've got to get this airplane moving. You sure you don't know where he is?"

Two days later I was in the office getting my gear together to go to South Africa when Gray finally showed up. The airplane was still parked outside, and, as Gray had suspected, Waldman wanted him to take it.

"I just got back, Phil. I have to take a few days off first and get some things done."

"But the distributor's in a rush for this airplane. He's been calling every day. So I told him you left yesterday."

"What'd you tell him that for? The airplane's still sitting outside," I said. Now whoever took the trip would have to hurry so that Globe Aero wouldn't lose its credibility. Waldman, however, didn't see it that way.

"Oh, that's just normal business. Gray'll leave in a day or so, and no one will know the difference."

"Gray's not going to leave in a day or so," Gray said, adamantly. "Gray doesn't even have landing clearance for the Philippines."

"You're not going to let a little thing like that stop you?" Waldman asked, making it sound almost like a challenge. "Walt and I used to fly there all the time, and we never had clearances."

“The Philippines wasn’t under martial law when you and Walt used to go there,” I said.

“Ferdinand Marcos might not like the idea of me just popping in unannounced. I’d feel a lot better if I had clearance before I left Lakeland.”

Waldman was a master at this, because all of a sudden it was no longer a question of whether Gray would take the trip, but *when* he would leave.

“Boy, you guys want everything to be easy,” he said. “Okay. How long’s it going to take you to get there? A week? Ten days? We should have the clearance any day now. Why don’t you leave tomorrow, and I’ll telex it to Honolulu as soon as it comes.”

When Gray still appeared hesitant, Waldman went in for the close. “You’re worrying for nothing. It’ll take you four days to get to Honolulu. I guarantee the clearance will be there before you are.”

It was just like Waldman to use logic. Waiting for clearances took time, and time, as he reminded us, was money—for Globe Aero, for the pilot, and for the distributor, who in this case was an old and valued customer. Finally, just to get him off his back, Gray left, with no clearance, but with Waldman’s assurance it would arrive within the next few days.

Gray wasn’t surprised when he landed in Honolulu three days later and there was no clearance. He waited four more days. When it didn’t come, he called Waldman and told him he was leaving that night and to telex the clearance to his next stop, Majuro. Again Waldman told him he was worrying needlessly and assured him the clearance would be waiting for him by the time he got there.

When Gray arrived at Majuro and then on Yap Island in Micronesia, and there was still no clearance waiting for him, he called Waldman, who assured him that he should have it tomorrow and he would send it to him immediately. This was the frustrating part of ferrying airplanes. When the airplane was broken, the weather was good; or when the weather was good, the airplane was broken; or like now, the airplane and weather were both good, but because of paperwork, the airplane couldn’t go anywhere. In any case, Gray would have to wait it out. With the situation in the Philippines, there was no telling how long it would take them to send the clearance, if they sent one at all.

Since the early 1970s, Marcos had been fighting the communists and the

Muslims, a crisis that led him to declare martial law, dissolve the congress, and arrest the opposition leaders. Marcos now ruled by absolute decree, and, while he didn't involve himself with minor affairs such as clearances for small airplanes, the military did. For them, inconveniencing the pilot of one small airplane was not of any great importance.

While Gray was on Yap, unbeknownst to him, Mantzoros, now flying for Piper Aircraft, was in Kota Kinabalu on the island of Borneo with a turboprop Piper Cheyenne that he and a factory mechanic named Francis Xavier Perullo were flying on a round-the-world demonstration flight. Like Gray, they, too, were waiting for clearance to fly to Manila. Unlike Gray, who was not getting paid while waiting, Mantzoros and Perullo, as employees of Piper, received their salary whether flying or sitting. For them, getting paid to lie in the sun on a South Pacific island was like a vacation—especially for Perullo, whose job at Piper offered few opportunities to travel outside the country, and he was determined to enjoy himself. "If these Filipinos wanna play games, who are we to argue?" Perullo said. "It wouldn't bother me to just sit here on this beach all day and drink beer. They'll call us when they want us."

Gray, on the other hand, wasn't particularly impatient, but like most of us, he had little tolerance for bureaucracy or small-time dictators. He got tired of waiting after one day, and, when the clearance still didn't arrive, he took off without it.

No one said anything when he entered Philippine airspace, nor when he landed in Manila. He went about the delivery like any other—detanking the airplane and packing everything in his suitcases. He headed for customs, thinking he was going to get away with it, when four soldiers stopped him and asked for his clearance.

"Oh, sure . . . the clearance," Gray said, smiling as he set his luggage down and pretended to search through his flight bag. "I've got it here somewhere," he said, while shuffling through flight plans, weather packages, and export declarations. "Is this it?" he asked, showing them the official-looking shippers declaration. They didn't appear interested, and Gray put it back in his bag. "Nope. Maybe it's in my other bag."

"Señor, we must have the clearance," said the soldier with the most stripes on his sleeve.

Gray went through the motions of searching through his luggage for another five minutes before giving up and disappointedly telling the sergeant. "I must've left it on the airplane."

The soldiers did not appear happy with Gray's revelation, least of all the sergeant, who was beginning to lose his patience and angrily asked, "Do you remember the number?"

Gray had no clearance number. But this soldier with the gun, and obviously the authority to make his life miserable, was insistent. Figuring he had better come up with something, Gray gave him the first number he could think of—his social security number.

"Wait here," the sergeant said, as he copied the number and then disappeared down the corridor, leaving the three other soldiers to make sure Gray stayed and waited. When the sergeant returned ten minutes later, he was even more agitated.

"This number is no good. You must have the right clearance number."

Feigning innocence, Gray kept to his story that the number he gave them had to be the right one. "I can't understand this," he said. "But look, the control tower wouldn't have let me land if I didn't have the clearance, right?"

The sergeant didn't answer; he turned to one of his men, said something in Filipino, and then turned to Gray. "All right. You come with me."

"Come with you where? I've gotta catch a plane." Gray said, no longer pretending disappointment.

"No, señor. I am afraid you will not catch your plane. You must follow me," said the sergeant, as he began walking down the corridor. The soldiers moved closer to Gray and led him to an office marked AIRPORT COMMANDER.

The commander, an army colonel with enough medals and braid to weigh down a person of lesser bulk, sat behind a massive desk and leaned forward while the sergeant spoke to him in Filipino, pointing at Gray and saying "Americano" and "permiso."

"Why do you come here without clearance?" the commander asked, in a humorless tone. "Do you think this is not serious?"

"I'm sure it's very serious, sir," said Gray. "But I had a clearance. I gave the number to the control tower. I just can't find it."

"But they do not have a record of your clearance. We have already checked. Do you think because you are Americano you can just land without permission?"

"No. Of course not . . ." Gray started to say.

"But you land anyway," said the commander, shaking his head and glaring at Gray. "This, I am afraid, is something we take very seriously."

"Okay, I made a mistake. What's the big deal? No harm was done."

"You have entered the country illegally," the commander said, raising his voice. "You have broken the law, and now you must pay for that."

"Okay, what is it? A fine?" Gray asked. Bribery and corruption, as he had heard, was how business was done in this part of the world. He could live with that.

"Yes, there will be a fine. But first we impound the airplane, and you, Americano, are under arrest."

Gray tried to protest, but the commander paid no attention. He spoke to the sergeant, who then escorted Gray outside, put him in a jeep with two soldiers, and told Gray they were taking him to jail.

As they drove past the main terminal, Gray happened to notice the familiar blue and white tail of the Pan Am Clipper parked on the ramp and figured it had to be the Honolulu flight. The same one he was supposed to be on. He checked his watch—thirty minutes before the flight was supposed to leave. There was enough time, but he would have to hurry.

"Stop!" he yelled at the soldiers. "Halt!"

The jeep jerked to a stop, and the soldiers turned to face Gray.

"Airplane," said Gray, and in part English, part Spanish, and mostly pointing and gestures, told the soldiers that all his papers were with Pan Am. "They arranged the clearance. If I could just talk to them this whole matter can be cleared up immediately."

"Pan Americano?" the driver said, looking at the airplane and then at Gray. "They have papers?"

"Si. Si. They've got all the paperwork. They'll straighten all this out if you just let me go talk to them."

If it was anyone else, they probably would've been silenced with a rifle butt and the jeep would've kept right on going. But not Gray. His boyish innocence seemed to invite trust. The soldiers exchanged a few words in Tagalog before turning to their prisoner. "Okay, señor. You get papers," said the driver, reluctantly. "But you must hurry. We wait here."

Promising that he would indeed hurry, Gray grabbed his suitcases, climbed out of the jeep, and then added for effect, "Now you stay right here because I don't want to lose you."

As soon as Gray was inside the terminal and out of sight of the soldiers, he hurried over to the Pan Am ticket counter and checked in for the flight. By the time the soldiers figured out their prisoner was not coming back, Gray was out of the country and well on his way to the safety of international waters.

Six hundred miles to the southwest on the island of Borneo, Mantzoros and Perullo were still waiting for landing clearance for Manila. They knew nothing of Gray's plight—if they had, they might've been suspicious when they received a telegram from Pat White, the distributor who was handling their demonstration flight, telling them that everything was set, and he'd meet them in Manila. When Mantzoros and Perullo landed in Manila, they were directed by the control tower to follow two army jeeps that pulled out in front of them.

"This doesn't look right," Mantzoros said, suspiciously.

"Hey, we were on Borneo for four days. Maybe war broke out. Maybe we're the enemy," Perullo joked.

As soon as he turned onto the parking apron and shut down the engines, the jeeps—now joined by two more—surrounded the airplane. Then, several soldiers armed with machine guns climbed out and stood by the airplane's door. "I've been to some pretty tense places, but I've never gotten a reception like this. Something's wrong," Mantzoros said.

"Well, I guess we gotta go talk to them," said Perullo, as he climbed out of his seat and walked toward the rear of the cabin.

"See if you can drag this out. I'll try to get someone on the radio and see if we can get hold of Pat White here, and he can straighten out all this stuff."

Perullo stood at the rear of the cabin, twirled up the ends of his handlebar mustache, and threw open the door. "Hellooo," he began. "How is everybody? Good, huh? Well, I'm kinda tired myself. Been flying all night, ya know."

The soldiers stared grim-faced but made no threatening moves, which encouraged Perullo to continue.

"Now where are my manners? My name's Francis Xavier Perullo."

The soldiers said nothing.

"Too much, huh? Okay, you can call me 'FX.' That's what all my friends call me. Except Sneaky."

More of nothing from the implacable soldiers.

"Who's Sneaky, you ask? Oh, she's a gal up in Lock Haven. You know, small town in the valley. Yeah. Well, that's where Sneaky lives. We've known each other for years, and just between you and me, lemme tell ya, I'd marry Sneaky in a minute if I could. What? Yeah, I know. Why don't I? Well, my friends, the problem is we never seem to be single at the same time. Either she's married when I'm divorced, or like now, she's single and I'm on my third marriage. Women. What're ya gonna do with 'em?"

While Perullo was rambling on and trying to stall the soldiers, Mantzoros was calling operations on the radio and trying to get someone to call Pat White. He gave up after twenty minutes and went back to check on Perullo, who, having exhausted the subject of women, was now jabbering away with stories about when he was in World War Two.

"What were you, about two years old during the war?" asked Mantzoros.

"Oh, something like that. But these guys are my buddies. I think they're finally coming around," Perullo said, while turning to face the soldiers again. "Remember those good old days? Boy we sure kicked the shit out of those Japanese, didn't we? I remember the time when . . ."

"Silencio!" one of the soldiers shouted. He took his gun from his shoulder and pointed it at the two pilots. The others did the same.

"Real good, FX. Now you did it," Mantzoros said, as both he and Perullo instinctively raised their hands in surrender.

"I can't understand it," Perullo said, looking hurt and confused. "I thought we were getting along great."

"Come! Quickly!" the soldier ordered, and ushered them into one of the jeeps. He took them to the terminal building and down a narrow corridor to an office with a sign on the door that read AIRPORT COMMANDER. The Filipino officer looked up from behind his desk as Mantzoros and Perullo entered the room and immediately accused them of violating Philippine airspace and entering the country illegally.

Mantzoros tried to protest, but the officer would not let him get a word in.

"This is getting to be a habit with your company," he said. "This is the second time you land here. I am afraid that this will not be tolerated."

"Wait a minute," Mantzoros cut in. "What do you mean, the second time? I've never been here before."

"You, no. But another pilot landed here without clearance only two days ago. That is why we let you land."

"Ah, general . . ." Perullo started to say.

"I am a colonel, not a general."

"Only a colonel?" Perullo asked, looking surprised. "Wow, you look important enough I thought you had to be at least a general."

"My rank is not what you should be concerned with right now, señor."

"Okay, colonel," Perullo said. "But listen, whoever landed here wasn't with us. We're out here for Piper Aircraft, and we're working with Pat White, the Piper distributor in Singapore. We don't know anything about another airplane."

"You admit then, that you work for the same company, a Señor White's from Singapore. And you are flying a Piper, just like him."

"Now hold on, colonel. Just because we're flying a Piper doesn't mean we work for the same company," Mantzoros said.

The colonel didn't hear, or he chose not to. All he could think about was that these two pilots would pay for that other American who tricked his soldiers while they were taking him to jail and sneaked out of the country. "You think you can land here without permission, just like your friend. Well, Americano, you cannot. And you will not escape like your friend, this, uh . . . Captain Gray."

"Gray!" Mantzoros exclaimed.

"Gray?" Perullo asked. "Who's Gray?"

"Dave Gray. I used to fly with the guy. All this goddamn trouble because of him." And then, apparently picturing his friend sneaking out of the country and making this commander and his soldiers look kind of silly, Mantzoros started to chuckle.

The commander, however, did not see the humor in what had happened. "We will see how funny you think it is when you are in jail. And you are not going to escape like your friend the spy."

"Spy? Gray?" queried Mantzoros.

"Whoa, whoa. Let's not go crazy here, your colonelship," Perullo said. "We had permission to land."

The airport commander started to say something but was interrupted by Mantzoros. "Look, we've got a telegram from Pat White. All you have to do is get ahold of him, and he'll clear everything up. He's got all the paperwork and all these other things. We're not going to listen to anymore of this shit," he said, turning toward the door. "We're getting out of here."

"You *are* going to listen to this shit!" shouted the colonel, as he slammed his fist on the desk and came out of his chair. "You are both going to jail, and you can rot there, Americano. You will not . . ."

"Uh, excuse me," Perullo said. "Can you just . . . can you stop this for a minute?"

"What?" the colonel demanded, agitated at being interrupted. "What is it now?"

"I gotta go to the bathroom. Can I go to the bathroom?"

The colonel started to say something, thought better of it, and let them both go to the bathroom with a soldier along to make sure they came back.

"We're in some big trouble here," Mantzoros said, when he and Perullo were alone in the men's room. "Boy, I'd like to get hold of Pat White, or we're gonna end up in the hoosegow."

On their way back to the commander's office they passed a bar, which reminded Perullo that he hadn't eaten or drunk anything but water since they left Borneo. "Hey, let's stop for a beer."

"I could sure use one after that circus. You think our friend here will let us?"

"Let's give it a try." He turned to the soldier. "Amigo. Stop for one beer?" he asked, pointing to the bar as he cupped his hand and raised it to his mouth.

"Cerveza. Una cerveza. Cinco minutos," Mantzoros said, and like Perullo, raised his hand to his mouth as if he were drinking.

The soldier looked at his watch, nodded, and allowed them five minutes before escorting them back through the terminal.

The commander lectured them for a few more minutes before sending two soldiers to put chains around the airplane's propellers so nobody could start the engines. "And you two," he said, pointing his finger at Mantzoros and Perullo, "are under arrest."

Having been tricked once, the commander was taking no chances this time. He sent Mantzoros and Perullo to a downtown hotel in the custody of two soldiers with strict orders to stand guard outside their room and under no circumstances to let the two Americans leave without first clearing it with him.

"Jail?" exclaimed Pat White, when Mantzoros reached him by phone that evening. "Oh, my!"

"Well, they're calling it house arrest," said Mantzoros, "and actually it's not too bad."

"I'm afraid they're still angry that Gray sneaked out of the country. So when I kept asking for clearance for *your* airplane, they set a trap. I'm terribly sorry about all this."

"These things happen. What can we do now, though?"

"Do? Why I imagine we'll have to let the diplomats try to get you released. In the meantime, I'll cancel the demonstration flight for tomorrow."

Mantzoros was familiar enough with how governments operated to know that if *they* got involved, two months from now he and Perullo would still be here. "Maybe you don't have to cancel the flight just yet. If you can pull some strings, we can still fly the airplane."

"That sounds bloody dangerous. After all, you chaps *are* under arrest."

"Yeah, but Perullo's goofing around with the guard and trying to soften him up. With any luck, we should be able to get out to the airport tomorrow."

Several seconds passed before the Englishman continued. "Perhaps there are a few people I can persuade to look the other way. I'll start working on getting the airplane back, and if you're sure you want to go through with this, try to be at the airport by eleven tomorrow."

The following morning, Mantzoros and Perullo bribed the soldier standing guard outside their room with one hundred dollars to let them leave the hotel, promising to be back in five hours. When they arrived at the airport, they parted with another hundred-dollar bill to get the soldiers watching the airplane to take the chains off the propellers and were in the air by eleven.

They flew to the southern airbase, took several military officers on the demonstration flight, and were back in the air and heading for Manila by midafternoon.

The freedom they were now enjoying prompted Perullo to suggest that maybe they should forget about returning to Manila and fly out of the country while they still could.

"That'd make a lot of people look pretty silly," Mantzoros said, chuckling as he pictured how the airport commander might take their sudden disappearance. "But that might not look too good for Piper. And I know Pat White would be in a lot of trouble."

"Yeah," agreed Perullo, though reluctantly. "They'd probably shoot those guards we bribed, too." So, as they promised, they flew back to Manila, locked up the airplane, put the chains back on the propellers, and returned to the hotel.

Two days later, Mantzoros and Perullo, along with Pat White and the U.S. Consul to the Philippines, made a formal apology to the airport commander, as well as a cadre of officials in the Philippine government. Publicly, of course, the apology was accepted, and Mantzoros and Perullo were allowed to leave. The United States and the Philippines were, after all, allies of a sort. Yet, what was never known—although widely speculated—was that the official, diplomatic apology also included a generous monetary offering.

14

The Dark Continent

Paul Crandall got out of the small Cessna after flying seventeen hours from Gander to Tenerife in the Canary Islands looking as if he just stepped out of the shower and dressed in freshly laundered clothes. His slacks had a razor-sharp crease, his shirt was clean and unrumpled, and every hair was neatly in place. Still, he ran a comb through his locks. After the usual "Welcome to Tenerife" and "Where are you going?" exchange, he could hold back no longer.

"Well, aren't you going to ask me about it?"

Crandall flew for Bob Iba's company out of Miami and though I had never met him before, I knew what he was referring to. It had been several weeks since the incident happened, and I doubt there was a ferry pilot who hadn't heard the story. I still pretended not to know what he was talking about, figuring I would have a little fun at the expense of knocking his ego down a peg or two. "Ask about what?" I said.

"The giraffe! Didn't you hear what happened?"

"I didn't hear anything. But I've been out in the Pacific a lot lately . . . over in Indonesia. You ever go there?"

"No, but have you ever been to Botswana? That's where it happened." He was dying to tell me, but I wanted to play this for all it was worth.

"I was there once, a few years ago, when I brought an airplane to a game reserve. I met this girl who worked in the office: blond, blue-eyed, and with this look that made me think of southern California. Then she started talking, and I knew it. So I asked her where she was from, and sure enough, she said Los Angeles. And here she was, in the middle of the bush, miles from nowhere."

"But did you see any animals, like maybe a giraffe?"

"I was with this surfer girl from southern California; I really wasn't paying attention to animals."

"Well, I saw one. Up close," he said, and finally blurted out, "In fact, I hit one."

"What do you mean *hit*?" I asked, figuring he had worked hard enough to get to this point.

"Hit. Like with an airplane."

"How do you hit a giraffe?" I asked, with just the right amount of surprise.

"Well, let me tell you," he began, happy now that he could tell the story one more time. "I'd left Sao Tome around sunset, flew all night, and was over the Kalahari Desert when the sun came up. I was trying to stay awake, but the sun was beating down on me, and I kept nodding off. So after a while, I dropped down over the desert and flew on the deck. That usually keeps me awake. I was just zipping along, having a great time, when I saw a herd of giraffes."

"You know, as many times as I've been to Africa, I've never seen any wild animals."

"Neither have I, so when I saw the herd, I got out my camera and started taking pictures. I guess I was about twenty-five feet off the deck, snapping away alongside the herd, when I slammed into something and crashed."

"Crashed?" I asked.

"Yeah. I thought I hit a tree or something. Anyway, I don't think the airplane rolled a hundred feet before it came to a stop. I didn't get hurt or anything, but, believe me, I was a little shook up."

"You're lucky it didn't catch on fire."

"I wasn't even thinking about fire. I just wanted to get out of the airplane.

But when I hit the ground, I must've buckled the fuselage, because the door was jammed. So I had to smash the Plexiglas windshield with the aircraft's towbar I had under the seat. When I finally got outside and looked at the airplane, then I was shook up. Here I was, in the middle of the Kalahari Desert with a wrecked airplane and a big dent in the wing. I still didn't know what I hit, so I followed the marks back to where the ground was all dug up from the crash. And lying there was the biggest giraffe I've ever seen in my life—and its head was gone."

"I gotta hand it to you. Anybody can hit a bird, but it takes a lot of skill to knock out a giraffe," I said.

"Yeah . . . well . . . now I had to figure out how to get out of there. I walked back to the airplane and, thankfully the radio was still working, so I put out an SOS to an airliner flying overhead and started waiting for search and rescue. A helicopter came out later that day, but it couldn't land because the weather turned lousy. They circled a couple of times, though, and dropped some supplies that I dragged back to the airplane. At least now I had food and water, something to survive on while waiting for the weather to clear."

"How long did you have to wait?" I hadn't heard this part of the story and was curious about how he was rescued.

"Man, it took them three days to get out to me. So every night, I slept in the airplane and listened to the wild animals tearing apart the giraffe."

"And you sitting there with a broken windshield a hundred feet away. Must build a lot of character," I said.

"I was just glad when the weather finally cleared up and a South African air force helicopter picked me up and took me to a military base in Gabarone. Now, all these doctors started checking me over to make sure I wasn't hurt. I kept telling them I just wanted to get something to eat, but they gave me a complete physical. Then they had me fill out a bunch of forms and kept asking me questions, like 'What was I doing in Botswana?' and 'Why was I flying so low?' "

"I imagine that was kind of hard to explain."

"I just told them I was tired. But they kept asking me more questions and giving me more forms. Finally, after about three hours, they let me scarf down some food and then flew me to Johannesburg."

"What'd the customer say about you wrecking his airplane?"

"He thought it was funny. Everybody in Jo'burg got a big kick out of it. Nobody'd ever hit a giraffe before. The press came to the hotel to interview me. I made the newspapers. They even had it on TV."

"You must be a celebrity down there."

"In South Africa, yeah. But not in Botswana. After all the crap I went through, they fined me for hunting without a license."

The trouble with Africa wasn't only that most of the legs were so long and monotonous that we often pulled some reckless stunt just to fight the boredom. It went much deeper than that. Africa was a political hodge-podge of tribalism, capitalism, and Marxism. Countries were paranoid about their airspace and not always friendly to pilots flying U.S. registered aircraft. No matter how careful we were, it was still risky—sneaking around squalid sub-Saharan cities and crossing borders on surreptitious flights over hostile airspace. All the while we hoped that we wouldn't end up on the wrong side of an old but deadly surface-to-air missile that would blast us out of the sky and scatter flaming debris over miles and miles of hopelessly dense jungle.

The trick to staying out of trouble in Africa was to keep a low profile. No place was easier for that than Abidjan. After landing, I would park in front of the control tower and walk toward the terminal as if I was going to the customs office. Once out of the control tower operator's line of sight, instead of going into the terminal, I walked along the outside of the building to the traffic office to pay the landing fee and check the weather. I then followed the same route back to the aircraft, figuring the tower operator, having seen me walk toward the terminal and then back to the aircraft fifteen minutes later, would assume I had gone through customs. Then all I had to do was taxi the aircraft to the Air Interivoir hangar at the general aviation section of the airport and slip into the country through the back door.

Although the police, as well as the American Embassy, probably wouldn't appreciate such flagrant disregard of local customs laws, it didn't stop me. Abidjan was too easy not to sneak into. The airport terminal was almost always crowded, keeping customs too busy to bother with a few harmless ferry pilots. So we tiptoed our way into the country and hoped that no one would ask to see the entry stamp in our passports, which, few of us had.

Of all the countries ferry pilots flew to in Africa, the one that caused us the most anxiety was Angola. As the country's independence drew nearer, conditions worsened. Yet, we continued flying into the capital during the riots, when the colonial government was falling apart, and right up to the end when the rebels were advancing on the city. It was during this time that Kerby and Wheeland were on a routine flight to South Africa and refueling their aircraft in Luanda.

A TAP Boeing 747 glided onto the runway with its position lights out—a darkened ghost of a ship illuminated in the night by a row of cabin windows like a string of pearls gliding to the ground. People were all over the airport, scrambling with what few belongings they could carry, to get out of the way of trucks careening across the ramp. They swarmed through the terminal, an uncontrolled and hysterical mob in a desperate hope of boarding one of the few remaining flights leaving the country. A sporadic thunder of gunshots and rockets, women and children screaming, and the squeal of rubber on asphalt all resounded as one.

A fuel truck jerked to a stop in front of Wheeland's airplane, discharging a Portuguese and an African who wasted no time refueling both aircraft. As the African rolled up the fuel hose, the Portuguese quickly scribbled out a receipt and urged the pilots to hurry. "The rebels are overtaking the airport! You must leave quickly!"

"We can't go yet," said Wheeland. "We have to go up to the control tower and file a flight . . ." The sharp echo of automatic weapons fire drowned out his words.

"Do not worry about that now," the Portuguese fueler warned, as he climbed back into his truck. "There is no one there to take it, anyway. Just get the hell out of here!"

Wheeland and Kerby taxied out to the runway amidst the pandemonium of rockets streaking across the sky and the whine of jet engines piercing above the din. A frantic confusion of airplanes scrambling in a mad rush to take off before it was too late. Wheeland, ever the disciplined, by-the-book pilot lined up on the runway and stopped to perform the last minute pretakeoff checks, when Kerby shouted over the radio, "Shooting stars are coming off your right wing!"

Not one to be easily excited or to use anything but textbook terminology over the radio, Wheeland quickly adapted to the situation when he saw tracers shooting above his aircraft.

"Holy shit! We gotta get outta here!" he exclaimed, as he opened the throttles and accelerated down the runway. Kerby took off right behind him.

Both aircraft were dark with no navigation lights to give away their position. Neither of them dared to say anything on the radio for fear the rebels could home in on their transmissions. They landed in Johannesburg just after sunrise with no flight plan and no prior warning—just two small airplanes that, according to Johannesburg Approach Control and the grateful insurance agent whose company had underwritten the ferry flight, were the last Western aircraft to get out of Luanda.

The communist-backed MPLA was now in power, and Soviet advisers came to aid their African comrades. UNITA rebels, supported by South Africa, were fighting a guerrilla war, but they were no match for the fifty thousand Cuban troops the Soviets brought with them.

So now we had to start looking for another country where we could get fuel. The problem with most countries on the west coast of Africa was either logistics or politics. Gabon was okay, but then you had to go way out over the water to stay clear of Angolan airspace. The Congo was in bed with Red China and hostile to Americans, and Zaire under Mobutu was too corrupt. I tried Sao Tome, a small, Portuguese island 150 miles off the coast, but they didn't have any fuel. This was something they didn't mention until after I had arrived.

"What do you mean, no fuel? Approach control told me you had fuel before I landed," I said to the refueler. By coincidence, I had arrived on Independence Day; the airport was crowded with locals, government officials, and delegations from neighboring countries. Everybody was celebrating, which is what this particular African probably wanted to do because he seemed in a hurry to get rid of me. I didn't intend on getting brushed off until I let him know how I felt about being told there was fuel and then finding none available.

Eventually, one of the African's buddies came over, followed by another and another. In a few minutes, there was a circle of Africans around us, and it was obvious whose side they were on.

"Whazza problem?" one of them growled. He looked straight at me, glaring, his face right up against mine. The railroad-track scars on his cheeks did an ugly dance. As situations go, this one did not look promising. I was an American in a country fed up with Americans and probably leaning to the East.

"Problem? There's no problem," I said, smiling. "Great Independence Day you're having."

I left as quickly as I could and flew to Libreville, refueled, and finished the trip before once again starting the search for some other country to fly to. The only place left was South West Africa, a former German colony bordering Angola to the north and South Africa to the east. At the end of World War One, the League of Nations gave South Africa a mandate to manage the territory's affairs, and so was presumed safe. Getting there, however, would not be easy. Between Abidjan in the Ivory Coast and Windhoek in South West Africa was more than twenty-three hundred miles of ocean with no navigation beacons and at a right angle to the tradewinds for a third of that distance. Worse yet, there was no way of knowing what the winds along the rest of the route were like.

The weather charts in Abidjan covered only as far south as the equator—four hundred miles at best. There were no forecasts or satellite pictures for the rest of the route, not even high-level charts that we could use to estimate what the winds were at lower altitudes. To most pilots, starting out on a long-distance flight with no weather briefing might seem foolhardy, and in just about any place else in the world it would be. But on the west coast of Africa, an almost stationary high-pressure area south of the equator dominated the route and favored us with clear skies for most of the flight. Before reaching this clear air, however, we had to fly through the intertropical conversion zone between the northern and southern hemispheres.

Because of the distance, I usually left Abidjan in late afternoon or early evening when this band of weather along the equator was at its worst. Huge air-mass thunderstorms climbed to stratospheric heights and could toss a small airplane around violently if it ventured too close. Great plumes of billowing clouds towered everywhere, and any thoughts of trying to hold a steady heading were quickly abandoned. All I could do was zigzag around the buildups until I reached the equator and punched through the last cloud.

The clear skies of the South Atlantic were home for the next seventeen hundred miles—a pitch-black sky filled with tiny pinpoints of light from a thousand stars. The aircraft's position lights were turned off so that no one could see me, and the only light for hundreds of miles was the dim red glow illuminating the instrument panel.

This was loneliness—the inability to see the ocean in the darkness, not receiving any navigation signals, and not talking with another human being until the approach at Windhoek twelve to sixteen hours away. My only companions were the lights from distant suns that traveled unimaginable light years to reach our terrestrial skies. What made it even lonelier was that only a few people in Abidjan knew where I was, and I had lied to them about my destination.

The problem with filing a phony flight plan was that if no one was expecting me, no one would come looking for me if I had to ditch. This added to my feelings of insignificance. It was as if I didn't exist. All connection with the planet had been severed—visually, aurally, and electronically. I might as well have been flying in a void, alone in the vast universe, and pass the time gazing up at the stars and searching for UFOs. If there was ever a time to see something that might not be of Earthly origin, I figured it would be between Abidjan and Windhoek, when I was divorced from everything terrestrial.

But the heavens never changed. No new lights flashed across the sky. I remained totally alone, preoccupied with the sedentary passing of the night until I estimated I was at seven degrees south latitude and had entered Angolan airspace. Then it was time to forget about alien spacecraft from another galaxy and turn my attention closer to home. In another hour I would be abeam the Soviet MiG base at Mocamedes, but far enough away that they wouldn't see me on radar—if I was on course.

I didn't know what the long-range capabilities of the radar installations in Angola were, and consequently, whether or not they could see me. But I knew they were watching, staring at their radarscopes all night, and waiting for some unidentified blip to appear. Then what would happen? Would they scramble fighters to intercept me? And once intercepted, would they force me to land or just blast me out of the sky right then and there and be done with me?

Ferry pilots' fear of flying close to Angola was legitimate for two reasons. The first was because of the U.S. registration on the aircraft, which easily gave away the country of origin. Since Angola had broken off diplomatic relations with the United States, any American pilot's presence there, if discovered, would not be looked upon favorably. The second reason was that most of the aircraft we ferried to Africa were being delivered to South Africa, the same country that was making forays deep into Angola in pursuit of SWAPO (South West African Peoples Organization) insurgents.

The Angolans didn't want us there, neither did the Russians or the Cubans. They might want the airplanes, but they had no use for the pilot, whose fate thus became uncertain. Unfortunately, for several hundred miles of this leg, Angola was the only place to land if a pilot did get into trouble. It was better than ditching at night in shark-infested waters with no hope of being rescued. I couldn't help thinking about this as I estimated my position to be crossing the fifteenth parallel south of the equator. The tradewinds had been blowing between ten and twenty-five knots, so I wasn't sure how close I was to the Angolan coast. I could be twenty miles from a squadron of Soviet MiGs or three hundred miles out to sea on a course to Antarctica.

Neither scenario offered a promising conclusion to the flight. While I tried everything from calculating the mathematics of estimated wind and airspeed to simply flipping a coin, I was never sure how much of an angle I should use to correct for this drift. So I worried about it until the sky began getting brighter with the approaching dawn and it was light enough to see that the isolated lights below me were houses and not ships. It meant once again that I had guessed right and would soon be on the ground. I had made several flights along this route and had guessed right every time—except once.

During that one time, as usual, I had been flying all night and figured I would cross the South West African coast just after sunrise. That should have happened two hours ago, but still there was no land in sight. To make matters worse, I had only three hours of reserve fuel. Even after making landfall, there weren't many airports along this section of the African coast. That was assuming I would be able to reach the coast. I had been holding a southeasterly heading, but now I began to wonder if my faith in the aircraft's compass had for the first time been misplaced.

It was time to forget about the compass and put my faith in something more certain—something that hadn't changed since the solar system was formed—the sun. I turned directly toward the rising yellow disc, sure this new direction would result in my intersecting some section of the coastline. There were no navigation stations along this part of the coast, and, with the sun still low, trying to get anything on the radio was a waste of time.

After an hour, I spied a darkened line on the horizon. It was too far away to tell if it was land, clouds, or just my imagination, but I knew that sooner or later I had to run into the continent. Crossing my fingers and praying, I aimed for what I hoped was the coastline. Twenty minutes later I was certain. It was land. I didn't know whose land it was, South West Africa or Angola, but at this point I didn't care. Anything was better than the ocean.

I was well into my reserve fuel when the coastline passed beneath me. The fuel-air mixture was already as lean as I could get it, but I leaned it again, hoping to squeeze every mile I could from what little fuel was left. The sun was high enough now that I was able to receive a good signal from the radio beacon at Rooikop, which only confirmed my fear. I was south of Angola all right, but I was also south of where I was supposed to be—a good two hundred miles south.

The left wing tank ran dry at one hundred thirty miles from Windhoek. I had used up all the fuel in the ferry tanks two hours ago and was now down to whatever fuel remained in the right wing—the last fuel tank. I figured there was about one hour's worth left, just over one hundred miles if I was lucky. I flew for another fifty minutes and was close enough to see the city, when the engine sucked the last drops of fuel out of the tank and the airplane turned into a glider. It was time to call for help.

"Windhoek Tower, from November-Three-Oscar-Hotel, I'm about ten miles west and just ran out of fuel. I think I'm going to set it down here for awhile," I said, trying to sound like some daredevil bush pilot who did this sort of thing on a regular basis.

The Afrikaner controller sounded just as calm. "This happens a lot over here," he said. "There should be a road beneath you. Can you see it?"

"Yes, I'm heading for it now. There're a few houses along it. I'm going to land close to one of them."

"Good. Try and get to a phone after you land and give us a call."

After listening to the steady hum of the engine all night, the silence was almost relaxing. I had been in the air twenty-one hours and, because there was no autopilot, I had had to fly the airplane by hand. Although tired, sleep was the furthest thing from my mind.

Circling over the road until I was at a low enough altitude, I turned onto a short final approach and made a last minute scan of the road for power lines, animals, or any other obstacles. Thankfully, I saw none. The road looked narrower as I got closer to the ground, but there was nothing I could do about it now.

The main wheels plopped onto the ground as I eased back on the control wheel, feeling the vibration in the rudder pedals when the tail wheel came into contact with the dirt road. The wings just cleared the bushes clustered on either side. I veered onto the shoulder and braked to a stop, got out, and pushed the airplane clear of the road. The air was still and all was quiet when I started walking toward the farmhouse I had noticed a quarter of a mile away.

I was in no particular hurry. Just glad to have landed safely, I was now enjoying the early morning crispness in the air when it hit me. This wasn't some cornfield in Iowa where one could just go for a leisurely stroll in the country. This was Africa. There were primitive bushmen here, SWAPO guerrillas, and wild animals in the bushes probably eyeing their breakfast. I quickly forgot about the easy stroll and ran for the farmhouse.

It looked deserted, no one milking cows or gathering eggs. There was not even a stray dog barking at me as I knocked on the front door. An old woman—white, with hard Teutonic features—answered and looked at me as if I'd just beamed down from another galaxy.

"Wie geht es Ihnen?" she said.

It sounded German, but it could just as well be Dutch or Afrikaans. I hoped it was German, because of the three languages, I knew more of German. I knew how to pick up women and tell people off, neither of which was appropriate at this time.

"Guten morgen. Ich sprechen nicht Deutsch," I said.

She understood that much, but looked puzzled and said something I didn't understand.

"Airplane," I said, and pointed toward the sky. "Flugzeug."

"Ja, das Flugzeug," she said, smiling and opening the door. I guess she understood airplane, but I didn't know if she thought I came from an airplane or I was looking for one. Because she didn't slam the door in my face, I figured I had her interest.

"Telephone," I said. I cupped my left hand and brought it to my mouth and my right finger to my ear and drew small circles. I must have looked like an idiot.

"Das Telefon, ja. Bitte kommen Sie herein," she said, while motioning me into the room. She showed me to the telephone and then said, "Darf ich Ihnen etwas Kaffee anbieten?"

I didn't know what she was saying, but it sounded like a question, which required a response. "Danke," I said, figuring whatever it was, thanks were in order. She left the room, leaving me to make my call, and returned several minutes later with a tray of coffee and more confusing conversation.

The fuel truck came a short while later, pumped thirty gallons into the wing tanks, and I took off for the short flight to the airport. Tomorrow I would fly to Johannesburg, deliver the airplane, and hand out the Oreo cookies, magazines, and Estée Lauder perfume I brought from the States. Every time I came to South Africa the secretaries and mechanics gave me a shopping list of hard-to-get items to bring back on my next trip. No matter the size of the airplane, I usually had enough contraband with me to ensure my popularity and celebrated return.

As for my contribution to international commerce, I too had a shopping list, which consisted of several bottles of South African wine. You couldn't get South African wine in the United States. You couldn't get gold Krugerrands or any other products made in South Africa because of the trade sanctions.

The United States and several European countries had ceased trading with South Africa in the hope that it would pressure the white government in Pretoria into changing its policy of apartheid. Diplomacy was at work. South Africans couldn't get Oreo cookies, but they could get airplanes. All they had to do was assure the U.S. State Department that the aircraft wouldn't be used for war, and they could get all the airplanes they wanted. The sanctions, however, did seem to be having an effect, because the signs for white and black restrooms were slowly disappearing.

While I didn't care for South Africa's racial policies, I liked coming here because it was an oasis and a safe haven at the end of a long, hard trip. It was a country where, unlike some parts of Africa, they didn't threaten to use my airplane for target practice. Politics aside, I liked South Africa because it was home to some of the most beautiful women on the planet. The words *stunning* and *ravishing,* although close, are inadequate in describing Afrikaner women because their features must surely be a gift from the goddesses themselves—a fact that becomes more apparent as one travels the breadth of this land. They become even more beautiful the closer one gets to Capetown, with its centuries of Dutch, English, Asian, Indian, and black African unions, a mixing of the races considered a blasphemous crime against society by the white government.

Most ferry pilots had a favorite city or country they liked to go to—some place that lured us like the sirens of ancient myth. Holmes liked Australia. With Kerby, it was the Pacific, especially Singapore and the islands of Micronesia. As for me, I liked Africa and went there so often that the other pilots began calling me the 'African Expert,' a title that implied I had mastered the dangers of its unpredictable weather, its archaic air-traffic system, its inefficiency and corruption, and its hyperventilating governments with their trigger-happy armies.

In reality, you didn't have to be an expert to fly in Africa. You had to be more like a short-order cook who knows a little bit about a lot of things. Besides knowing which airports had fuel, which ones were friendly, and which ones charged a hefty runway-light fee, you had to know about the thunderstorms that always hung over the southern Sahara, and that they were weakest in the early morning. Likewise there are the buildups over the equator. You had to know that if you tipped the refuelers in Abidjan or Dakar a pack of cigarettes or a *Playboy* magazine, they wouldn't take all afternoon to refuel your airplane. Neither did it hurt to have some knowledge of local customs—which, according to rumor, Holmes did not.

As the story went, every time Holmes landed in Abidjan the same African refueled his airplane, and Holmes always thanked him by forming a circle with his thumb and index finger as if to say "Good job." Every time he did this the African became angry, finished refueling, and went storming off in the fuel truck, leaving Holmes to wonder what he did to insult the man so grievously. Holmes made several trips to Africa before learning that the gesture he had been making, while perfectly acceptable in the United States or Europe, to an African represented a hole, specifically an unclean hole in one's lower anatomy.

Although I readily briefed other pilots who were flying to the continent, I tried to keep all the African trips to myself. Whenever I found out we had an airplane for Africa, I tried to position myself so I would be the only one there to take it. The strategy worked until we started getting too many airplanes for Africa, and Waldman wanted more of his pilots checked out on the continent. Because I had flown there more than anyone else with the company, the job of training these pilots fell upon me.

Globe Aero usually had only one airplane at a time going to Africa. Two airplanes, while infrequent, did occur often enough for me to train a few pilots, but only one at a time. When we were contracted to ferry three airplanes to the Nigerian air force in Kano, I was assigned to train two pilots, Tim Peltz and Bob Moriarty.

Peltz was a curly, blond-headed kid from San Francisco who looked as if he should have been on a college campus rooting for the next big game. Instead, he was on his way to Africa with me and Moriarty, an ex-Marine Corp fighter pilot and forward air controller who had been decorated in Vietnam. Moriarty had the rascally personality of Yosemite Sam and was a good ferry pilot who let everyone in on this little-known fact before ferrying his first airplane. He had been with Globe Aero just under a year, during which time we learned that what he lacked in ocean-flying experience, he made up for in intelligence and, at times, just plain guts.

It was just after sunset as we prepared to leave Gander. Snow was falling, and there was a light covering on the wings of all three aircraft. Knowing the consequences of taking off with even a thin layer of snow, Peltz and I began sweeping the wings clean with a broom borrowed from the maintenance shop. Moriarty, however, had other ideas.

"This stuff will come off as soon as I get on top," he said, as he looked up at the low ceiling. "It doesn't look that bad. I'm gonna get going."

"What, do you have a date?" asked Peltz. "It'll only take a few minutes to sweep off your wings. I'll even do it for you."

"You guys do what you want, but I'm getting out of here," Moriarty said.

He then got in his airplane and left. Peltz and I swept the snow from our aircraft before taking off one hour after Moriarty.

"You think he might be in trouble?" Peltz asked, after we had been in the air several hours and hadn't heard from him.

"Center would've told us if he went down," I said. "More likely he picked up some ice and is wafting through the air right now. It'll take a long time to get rid of with these temperatures, and, knowing Moriarty, he's probably too pissed off to talk to anybody."

We didn't hear from Moriarty for the rest of the flight, which was not unusual. Because of weather conditions, we could be as little as one hundred miles apart and still not be able to talk to each other.

Peltz and I landed at Tenerife late the following afternoon and checked into the small airport hotel adjacent to the terminal. There were plusher places we could've gone to—big, five-star resort hotels on the beach where the tourists stayed. The Canary Islands, just off the Northwest African coast, were a popular vacation spot for Germans, the English, Scandinavians, and the French. All came to escape the cold European winters and lie in the sun unrestrained by modesty, thanks to the Spanish government's liberal attitude toward topless sunbathing.

For ferry pilots, Tenerife was usually nothing more than a quick stopover to grab a few hours of sleep before continuing on to Africa. Looking for a fancy hotel seemed a waste of time, especially when the airport hotel was so much more convenient. It wasn't much, only ten small rooms; but there were always one or two available.

We were about to crawl into our bunks when Moriarty came storming into the room, threw his suitcase onto the third bed, grumbled something about having to land at St. Johns because his heater quit, and went right to sleep. Moriarty could fall asleep faster than anyone. I don't know about Peltz, but it took me a while to drift off. Then I was out until the late night wake-up call roused us from a sound sleep.

I always got a strange feeling when I was getting ready to leave Tenerife. Was it loneliness? Was it apprehension about the long flight? No matter how many times I came here, it was always the same. It was night, like now, and the gloomy walkway connecting the hotel to the terminal was deserted. Although tonight I had only one more leg left to fly, instead of the entire length of the African continent, and even though I had company, I still felt the apprehension. Moriarty and Peltz must have felt it, too, because they were just as quiet.

My mood worsened after seeing the eight crated jet engines stacked in the corner of the parking apron. The engines were all that was left from one of the worst disasters in aviation history, which occurred a few months earlier when a Pan Am Boeing 747 collided with a KLM 747, killing 582 people.

While we didn't have fog to contend with this evening, there was a low overcast that reduced the visibility to a few miles and obscured the twelve-thousand-foot mountain south of the airport. As we taxied to the departure end of the northwest runway, Peltz, in front of Moriarty and me, was the first to call the control tower for his departure clearance.

"Climb straight out on runway heading to five thousand feet. Turn right to a heading of one-two-zero degrees on course," the Spanish controller instructed.

As was standard procedure, Peltz read the clearance back to the controller word for word—except for one small detail. Instead of turn right, he said turn left. The controller must have missed the error, because he acknowledged Peltz's recital with a rapid, "Readback correct. Cleared for takeoff."

I had taken off from this same runway enough times in clear weather to know that a left turn would take Peltz south of the airport and straight into the side of the mountain. A mountain he couldn't see because of the clouds. I had to warn him, but the controller was talking to an approaching Iberian flight and tying up the frequency. Peltz was just lifting off the runway before I was able to get through.

"Make that a right turn, Peltz," I said into the mic, quickly, before anyone else could cut in. "Do you copy? Turn right."

Several seconds passed. Peltz was airborne now and almost in the overcast, before he finally answered, confirming my correction and adding a grateful, "Thanks." There wasn't much more he could say or that was even expected, because, had our positions been reversed, he would have done the same.

When all three of us were at cruise altitude, we joined up in a loose V formation. Although I usually flew in Africa with my lights out, we kept them on tonight so we could see each other. Africa had been quiet lately, which to us meant no aircraft had gotten shot down and no pilots had been thrown in jail or executed. Still, we decided to fly directly to Nigeria and not talk to anyone until reaching our destination. As we approached the African coast, heading for El Auin in the Spanish Sahara, I began thinking that flying over the capital city while there was a war going on probably wasn't the wisest thing to do. I confided my apprehensions to Peltz and Moriarty.

"Who's fighting who?" Peltz asked.

"Morocco and Mauritania," I told him. "They're both claiming a piece of the Spanish Sahara and are fighting the nationalists for control. It looks like they've been at it since the Spanish pulled out."

"Sounds like Angola," Moriarty said. "Welcome to Africa; hundreds of years of tradition unhampered by progress. Isn't independence great?"

"I've been thinking, since we've got our lights on, maybe we should call El Auin and let them know who we are," I said. "If they're watching us on radar, they might get nervous about seeing three aircraft in a military V formation heading straight for their capital."

"Why don't we just turn our lights off," Moriarty suggested. "If they're having a revolution, they might tell us to stay out of their airspace. Or worse, they might tell us to land."

"I don't want to land," said Peltz. "I just opened up a can of Beanie Weenies."

"You've got Beanie Weenies?" exclaimed Moriarty, which set off a lengthy assessment of V-8 and Screaming Yellow Zonkers.

Our conversation ended abruptly when Peltz curiously asked, "Hey, what are those lights up ahead?"

"Lights? Where?" asked Moriarty.

"Oh, Christ! That's El Auin!" I said, as I peered out the windshield and saw the lights getting closer. I had forgotten all about the city and how fast we were closing in on it. "I guess we better talk to them," I said, and switched over to approach control.

"El Auin, this is Five-November-Alpha-Zulu-Whiskey, flight of three civilian aircraft fifty miles northwest at one-one-thousand feet en route Kano, Nigeria. ETA El Auin at zero-one-two-two Greenwich," I said, with urgency.

Several seconds passed before a Spanish accented voice answered. "Five-November-Alpha-Zulu-Whiskey, from El Auin. It is good you call. We were ready to launch fighters."

"You don't need to do that," I quickly answered, "we're a flight of three unarmed civilian aircraft."

"Okay, okay," the controller said, wearily. "You call overhead El Auin."

I waited a few seconds and then said the single word "Company" to let Peltz and Moriarty know I wanted to talk with them on a discrete frequency.

"They gave in too easy," I said, when Peltz and Moriarty called in on the new frequency. "I don't trust them. I think we should go dark. We better change altitudes, too, so we don't run into each other. Peltz, you take eleven-five, and Moriarty why don't you drop down to ten-five. I'll stay here at eleven."

When El Auin was safely behind us, we switched back to our private frequency, turned our lights on and rejoined into a V formation. Peltz was the first to voice the relief all of us felt. "Maybe it was just a bluff, but I'm glad we didn't have to find out the hard way."

Moriarty, ever critical, couldn't decide what the hard way was. "I don't know what would've been worse, getting shot down or thrown in some Spanish jail."

"I don't know about getting shot down, but I doubt whether jail would be much better," I said. "I spent the night in Villa Cisneros once and wasn't too impressed. And that was in a hotel. I hate to think what their prisons are like."

The rest of the night was quiet. We skirted around a few buildups near Bamako, but most of the weather was west of us. The sun had been up for an hour when we landed at Kano. Although the delivery went smoothly, thanks to the Piper representative who had flown in from his native Switzerland for the occasion, we didn't finish until late afternoon. With the Swiss rep's help, we booked a suite at the Kano Intercontinental. When we arrived at the hotel, however, the large suite with three beds that we were promised, turned out to be a small, single room with two beds and one cot. Tired from flying all night and being up all day, we resigned ourselves to our lot, had a late dinner, and went straight to bed.

I couldn't have been asleep very long when I was awakened by someone tugging at my sheet. I turned over, sat up, and squinted my eyes to see who the shadowy figure standing at the foot of my bed was. It took several seconds to adjust to the darkness and make out the intruder. It was Moriarty, and he was clutching the corner of my sheet.

"Whad'ya want?" I asked, from some groggy netherworld between sleep and wakefulness.

"I need your sheet," he announced, matter of factly.

"Well, go take Peltz's. I'm using this one," I said, and turned over to go back to sleep.

The next thing I heard was the door opening and closing. I turned over to see what was happening now, but Moriarty was gone. I looked around the room for other clues, but saw only Peltz lying motionless in a darkened pile of pillows and blankets. Tired as I was and only half awake, I decided that this was a mystery I didn't care to pursue. I turned over, pulled the thin blanket over my head, and went back to sleep.

Some time later, I was awakened by a loud knock on the door. I sat up, rubbing the sleep from my eyes, and asked, "Who's there?"

"Security," said a slurred African voice from the other side of the door.

I looked at my watch, which showed one o'clock in the morning, and wondered what a security guard wanted at this hour, as I climbed out of bed and opened the door.

A uniformed Nigerian security guard with an automatic rifle slung over his shoulder stood in the hallway. Directly behind him, with his head down as if in humble repentance, was Moriarty. He was naked except for a pair of baggy white boxer shorts and several gold chains around his neck.

"Does this man belong to you?" the guard asked, pointing to the sorry figure of Moriarty, still looking down at the floor and not saying a word.

"Yeah, he's with us," I admitted.

Stepping aside and motioning for Moriarty to go into the room, the guard then turned to me and issued the stern warning, "See that he stays here. There is a curfew on."

I let Moriarty into the room, locked the door, and turned to ask what he was doing walking around downtown Kano in his underwear at one o'clock in the morning. But I was too late; he was already in bed and fast asleep.

Still curious the following morning, I asked Moriarty what happened last night.

"What do you mean?" he asked.

I told him about the security guard and about him trying to take my sheet.

"Now what would I want with your sheet?" Then he gave me a strange look, as if I was the one imagining things. But I knew I wasn't. I looked to Peltz for corroboration and asked if he had seen anything.

"I was asleep from the time my head hit the pillow 'til I woke up this morning," Peltz said.

No help there. I was sure it wasn't a dream. It was too vivid. I pressed Moriarty further until he finally confessed to some rumor about his alleged sleepwalking.

"Sleepwalking? In Africa?" Peltz exclaimed. "Where'd you go?"

"I don't know," Moriarty innocently remarked. "For a walk, I guess."

"I wonder what that security guard thought when he saw some crazy American walking around in the middle of the night in just his underwear," remarked Peltz.

"And after curfew, too. You're lucky you didn't get mugged," I said.

Moriarty shrugged off our concern over his brush with Nigerian crime. "That's okay," he said, forgetting about the gold chain he had around his neck. "I wasn't wearing anything except my skivvies, so they wouldn't have gotten much."

Moriarty was as restless during his waking hours as he was when asleep. Taking a break from ferrying, he prospected for gold in Alaska, and when that adventure proved to be more work than reward, he got involved in his first air race from Paris to New York. Finding the strategy needed for this type of flying to be more challenging, he entered a Paris to Libreville race, but got only as far as Portugal when he had to abort because his airplane was using too much oil. By the time he returned to Paris and had the oil problem fixed, however, it was too late to finish the race.

Not one to let this prevent him from getting at least some recognition for his efforts, Moriarty prepared for his next stunt by calling a French news agency. The following morning, after taking off from Le Bourget and before setting course for Shannon, he brought his aircraft in low over the city. He dipped down to treetop level and slid over the shoulder of the Ecole Militaire. Lining up his aircraft on the grassy pathway and in a final arc of precision that had not been attempted since World War Two, he swooped underneath the Eiffel Tower.

Moriarty got his recognition all right. He not only made the French newspapers, but a few American ones as well. *People* magazine ran a picture of him flying underneath the famous tower. He was a celebrity on both sides of the Atlantic, a distinction that would not help years later when he got into more trouble and what he needed more than anything was someone to vouch for his character.

15

Adrift in the Kingdom

The trip had gone well so far, giving no indication that our flight to Dar es Salaam, Tanzania, would be anything but routine. The North Atlantic was entering its pattern of winter weather, but the Navajo I was flying and Gray's Turbo Aztec easily flew above the clouds on the great-circle track from Gander to London. After dropping the aircraft off at the maintenance facility outside London at Gatwick for some minor refitting, we checked into a hotel and called Globe Aero, but they hadn't received anything from Khartoum yet.

Gray and I were not surprised. North Africa and the Middle East were always a problem when it came to clearances. Every single country wanted to know weeks in advance when we were going to enter their airspace and where we intended to land. Then all they gave us was a seventy-two-hour slot within which to comply before the clearance became void. It was worse when we wanted to fly through several countries, because every clearance had to be coordinated so that each three-day window fell into precise chronological order.

The responsibility for requesting these clearances had recently been assumed by Dave McCollum, for which we were all grateful for two reasons.

For one, McCollum tended to be more diligent in chasing after and coordinating the various clearances, and, two, his name affixed to a telex sent to Arab countries usually got a quicker and more accommodating reply than the obviously Jewish Waldman.

Not even McCollum was having much success in getting the response that Gray and I were waiting for. Three days later, as we were preparing to leave England, I called Globe Aero again, but still they had received no word from Khartoum.

"I don't believe it. This is the same thing that happened in the Philippines," Gray said, disappointed that he wouldn't make it home for Thanksgiving, now a week and a half away. "I knew I shouldn't have let you talk me into taking this trip."

Because we had reservations at the Athens Hilton, we decided to fly to Athens and wait there for the clearance. That put us two thousand miles closer to Tanzania, and it was the last stop before Khartoum. I told Waldman of our decision, adding, "But that's it. If the clearance doesn't come in by then, we're not going any farther."

We spent three more days in Athens, and still no clearance came. Globe Aero had received nothing from Khartoum—no approval, no denial, not even confirmation that they had gotten our request and were working on it. Gray and I had been in situations like this before, and, although we may have sworn never to get suckered in again, somehow we always did. And we always got out of it the same way out.

"I guess if we want to go anywhere it's going to be up to us," I said, when we sat down to an early dinner in the Hilton's outdoor café to figure out what we should do next.

"Okay, so let's think about our options," said Gray. "Our Egyptian clearance is only good for one more day. I guess we could wait here and get McCollum to request new clearances. But that could take another week or two, and we still might not get clearance for the Sudan."

"Are you thinking what I'm thinking?" I asked.

"Moriarty does it all the time when he doesn't have a clearance, and he seems to be getting away with it."

"But for how long? One of these days they're going to catch on to his tricks. I don't mind getting in trouble, but I wouldn't be too happy about

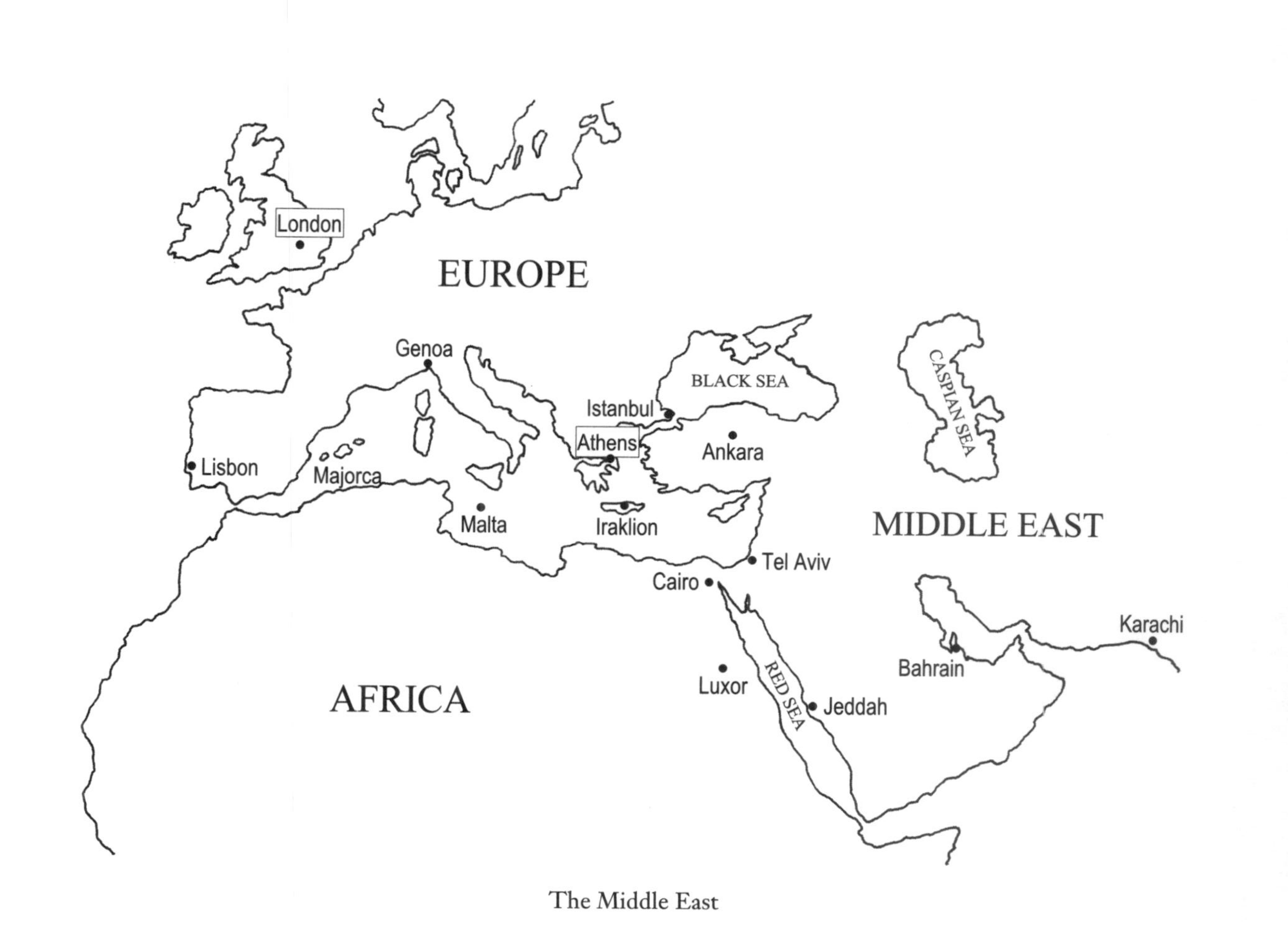

The Middle East

getting shot down or thrown in jail because someone was pissed off at Moriarty and decided to take it out on us."

"You've been there at night. Has anyone ever asked you for a clearance?"

"There's usually just some buck private in charge who only wants me to get out of there so he can go back to sleep."

"Same here. I figure if we left at seven tonight, we could be in Khartoum around three in the morning, refuel, and take off right away."

"You're probably right," I agreed. "They probably won't even figure out we don't have a clearance until after we leave."

Cairo, the great city of North Africa, passed by my left wing as no more than a dim blur of light filtered through the dirty haze of the desert. Not far to the east, the Nile River snaked an ancient course toward Khartoum and beyond. There was little life below us, even if we *could* see through the haze. The desert was asleep, and the only movement was the wind blowing at a right angle to our course and reducing the art of dead reckoning to no more than a guess.

The charts warned in bold type that strong action would be taken against aircraft violating the wrong airspace. Egypt was okay, as long as we had a clearance. So, too, was the Sudan, under the same conditions. The danger would be when we got close to the Ugandan border. If we were off course no more than five degrees in the wrong direction, all we could hope for was that Idi Amin, Uganda's self-proclaimed president for life, wasn't having a bad day.

Gray and I had been following Amin's rise to power ever since he tried to settle his country's border war with Kenya by challenging President Jomo Kenyatta to a boxing match. When no response came, the Ugandan president offered to fight with one hand tied behind his back. Still, the challenge went unanswered, the war escalated, and Amin's reign of terror grew increasingly bloody. Because Globe Aero ferried aircraft along Africa's east coast on a semiregular basis, we figured it was only a matter of time before one of us strayed into Ugandan airspace and was forced to land. Naturally, we went to great lengths to hone our navigation skills when flying in that part of Africa; Gray, however, took a more diplomatic approach.

"I figured it'd be a good idea to have him on our side. So I wrote him a letter saying how much we admired him and that we made him an honorary ferry pilot. I even sent him a Globe Aero sticker."

"The guy's a nut about uniforms. You should've sent him a medal, too."

"Oh, I sent him something for his uniform. The next time you see his picture, take a close look at his chest, because I sent him a little set of pilots wings."

My amusement over Gray's solution to a possible problem was cut short when, at fifty miles from the Sudan border, Cairo Radio called to advise us that Khartoum had forbid us to enter Sudan airspace.

I listened to the static hissing through the speaker, before the controller's next words indicated a direct response was required. "What are your intentions?" he asked.

I had no idea what our intentions were and told him I would get back to him in a few minutes. I then called Gray on our discrete frequency. "Did you hear that?" I asked.

"Yeah, I heard it. So now what do we do?" It was a rhetorical question, because I'm sure he knew I had no idea what to do next any more than he did.

"Maybe we can fly down the Red Sea and then cut across Ethiopia," I suggested.

"That's the same route the Israelis flew when they rescued the hostages in Entebbe. Everybody probably gets nervous now when they see airplanes flying down the Red Sea."

"Then we'll have to fly around it," I said, as I unfolded the chart and looked for borders and airways. "But we're going to have to land someplace and take on more fuel."

Luxor was the logical place. We didn't have clearance to land there or anywhere else in Egypt for that matter; all we had was overflight permission. Luckily, the controller understood our predicament, as did the Egyptians at Luxor. They fueled our aircraft as soon as we landed and were patient as Gray and I planned our next move.

"It looks like we're going to have to fly over Saudi Arabia, and down over Yemen, then around the horn here," I said, pointing out a route that would take us around Ethiopia and Somalia. "When we get abeam Nairobi, we can head inland, straight for Dar es Salaam."

Gray looked at the chart, obviously unhappy over the thousand extra miles and the countries we would have to fly over. "I guess we should be okay if we stay over water. But we're going to need clearance to land in Jeddah and probably Yemen, too."

"I don't think we're going to be able to get much done here, though. Probably the best thing to do is fly to Jeddah, take on fuel, and get them to help us get a clearance for Yemen," I said. It wasn't the best plan; it was the only one. Resigned to finish the trip in spite of the crazy routing, we had the control tower operator call Cairo Radio and request clearance from Jeddah. Thirty minutes later, they called back and told us the Saudis okayed our clearance.

Sunrise was little more than an hour old when we landed in Jeddah, yet already it was hot and muggy and, we were surprised to find, extremely busy. Airliners were taking off and landing every few minutes. Fuel trucks, buses, caravans of baggage carts, and an occasional black limousine as long as a Greyhound bus all scurried across the tarmac. We began the wait, yet no one came. We called the control tower to ask if someone could give us a ride, but fifteen minutes later when no one showed up, we set off on foot to find the traffic office.

Finding a door marked with a large C, we walked in and told the Arab behind the counter that we wanted to refuel our aircraft, pay our landing fee, check on getting a clearance to overfly Yemen, and that we would leave as soon as possible.

The Arab clerk looked at us wearily, glanced at some papers on the counter, then back to Gray and myself. "Yes. Yes," he said, impatiently. "You may refuel and file your flight plan, but first you must pay the fine."

"What fine?" Gray and I asked in unison, looking to each other and then the Arab for some sign of understanding.

"You landed without clearance," he said, as if this was an unpardonable sin.

We told him Cairo had gotten clearance for us, and all he had to do was check with air traffic control.

"Yes, but that is not the proper clearance."

Sure it is, we insisted, repeating it several times before figuring out he was referring to a diplomatic clearance, something entirely different from the air traffic clearance we had. Rather than get into a long argument about interpretation, we decided to just pay the fine. It couldn't be much, anyway, and if it made this one clerk happy, it would be worth it.

"The fine," he said officiously, and apparently pleased with his arithmetic, "comes to one thousand five hundred U.S. dollars."

That worked out to seven hundred fifty dollars per airplane. It was a lot of money, but Gray and I figured we could afford it, although we would have to scrimp for the rest of the trip. We counted out the money, but the clerk shook his head and wouldn't accept it.

"No. It is one thousand five hundred dollars for each airplane."

There are degrees of guilt and degrees of punishment. Though three thousand dollars might not seem like a lot of money to someone whose country is sitting on a hundred years' reserves of oil, to Gray and me it far exceeded the degree of punishment we felt was warranted. Trying to convince the clerk that our crime was not worthy of such harsh punishment, however, was not so easy. He stood behind his arithmetic and wouldn't budge.

"We'd like to talk to your supervisor or whoever's in charge," I said.

I might as well have asked to speak to Allah himself, for he just looked at us, quizzically at first, as if unprepared for someone to question his authority. He put his pen back in his shirt pocket, looked us squarely in the eye, and informed us that *he* was in charge.

"No, we want to speak with your boss, the person in charge of the airport."

By the look he gave us, one might think we had defiled his mother, his father, and all his ancestors. He stood firm, and defiantly parroted, "Yes, well, I am in charge."

Gray and I gave up. Maybe tomorrow, after a night's sleep, a shower, and a meal, things would look better. We told the clerk that we would be back in the morning to straighten everything out and went to the passenger terminal to find a telephone.

"Why's the place so crowded?" wondered Gray, as we fought our way through the crowd. "Something must be going on."

"Whatever it is, it must be big. I just hope we'll be able to get a hotel room."

We couldn't. We called one hotel after another, and at each one we got the same response—they were booked solid for the next two weeks. Our spirits were sinking. We began contemplating the idea of sleeping in our airplanes, when we noticed the British Airways 747 parked outside the terminal.

"If British Airways flies into Jeddah, then they must have a hotel where they contract rooms for their crews and passengers," said Gray. "Maybe we can talk them into getting us a room. Hell, we fly on them enough, they owe us at least that much."

The British Airways operations officer thought we were joking. "Rooms? Hah," he snorted. "This is the Haj. All our rooms are taken."

"You gotta have something," I said. "We'll have to sleep in our airplanes if you can't help us."

The overworked ops officer was sympathetic to our plight, but unfortunately, he told us, there was nothing he could do. With nowhere else to turn, Gray and I went into our routine. We pleaded. We begged. We told him we had been flying on British Airways since we were little kids and that we would fly on British Airways, and only British Airways, until we were old men. Finally, giving in to our desperation, he made a few phone calls and handed us a slip of paper with the name of a hotel scribbled on it. We thanked the operations officer profusely and walked outside the terminal into mass chaos.

It was the first time either Gray or I had been in Saudi Arabia, and we couldn't have picked a worse time. The Haj, the holiest of Muslim religious observances, was just getting underway, and all of the third world, it seemed, was going to attend. There were hordes of people of almost every description: young and old; infants whose tiny heads peeked out through their blankets; and the ancient ones, hobbling along as fast as their skinny legs carried them. They were sprawled on the sidewalks, sleeping in makeshift shelters propped up against the terminal building and huddled around cooking fires in the empty lot across the street.

A cornucopia of sounds and smells filled the hot, desert air, horns bleated their automotive impatience, and the burning aroma of alien foods mixed in the rancid air with diesel fuel belching from the exhausts of an endless stream of buses. Added to this attack on our olfactory senses was the stench of hundreds of unwashed bodies that had traveled on countless boats, airplanes, trucks, and maybe even the odd camel or two to their annual pilgrimage to Mecca.

Gray and I, two infidels in the middle of it all, took a taxi headlong into the bumper-car pandemonium of traffic. The airport was at the edge of the city, so close that there was no respite from the hysteria of transient mobs of people. There were four large dormitorylike buildings adjacent to the airport that the taxi driver told us were empty all year long except during the Haj. Then they were teeming with pilgrims. When those buildings were filled, the overflow was left to fend for themselves in the growing tent city in the empty lot across from the airport.

We followed the push of vehicles into the heart of the city, past ornate mosques and modern five-star hotels, which were a testament to the kingdom's newfound wealth in petro-dollars. While our hotel was neither large nor modern, it was nonetheless clean and well kept. The desk clerk, an Arab with a distinct British accent, registered us for one night only and at an exorbitantly inflated rate. At this point Gray and I would have paid anything. We still didn't know what we would be doing or where we would be doing it tomorrow, but for tonight, at least, we had a place to sleep.

While checking out of the hotel the following morning, the desk clerk informed us that we could have the room for another night. However, the rate would be double what we just paid.

"That's usury," I said, angered at such a blatant rip-off and surprised that he could charge it with such a straight face. I had always thought usury was against the teachings of the Koran, but obviously that credo didn't extend to this particular Arab.

"No, sirs, this is not usury," he said, not in the least offended by my observation. "This is business."

Suspecting that before this trip was over, we would probably need every dollar we had, we suggested to the desk clerk what he might do with his room, and left.

The airport was much the same as yesterday. Airplanes were landing every few minutes to disgorge their human cargoes, the terminal was still a confused mob of people, and the tent city was growing. The same Arab was running the show from behind his counter in the traffic office, and his indifference to our problem had not changed.

"I am in charge," he informed us when we asked to speak to the person in charge. He wouldn't budge on the fine and then told us we couldn't fly from

Saudi Arabia to Yemen as we had hoped because of a border conflict between the two countries.

"Yes, of course you're in charge," I said. "But is there anyone else we can talk to about helping us get a clearance to overfly Ethiopia or anywhere?"

"No, there is no one else. I am in charge."

We were becoming more frustrated every time we talked to the clerk. The situation wasn't hopeless, but it appeared to be leaning in that direction. We couldn't leave until we paid the fine, which, even if we did, we still couldn't leave because we had nowhere to go. We were running out of ideas when I suggested this might be a good time to call for help.

"You mean like get Kerby to plan a rescue mission?" Gray said, jokingly.

"We may have to do that yet. But right now I think we need to call the American Embassy."

Like most Americans, I've taken my share of potshots at the U.S. government. I've criticized its often inane policies, questioned the intelligence of its politicians, and vowed on more than one occasion to move to Andorra and burn my American passport. But for all my complaining, I was never happier than when the taxi dropped us off in front of the embassy with the Stars and Stripes waving in the breeze.

From the ramrod-straight bearing and barrel chest of the young marine guard who escorted us into the main building to the efficient, yet friendly and matronly receptionist, I felt we had finally found an ally. She offered us American coffee before bringing us in to see the assistant ambassador, a tall, distinguished man in his mid-fifties who spoke with the careful voice of a diplomat. He listened as Gray and I recounted the events of the last two days.

"It sounds like you fellows have certainly had a difficult time, but what is it you'd like us to do?" he asked.

"We were hoping you could use your influence with the Saudi government to waive this fine and help us get a clearance," I said.

"Let me explain our position here. The State Department doesn't have a lot of influence with the Saudi Arabian government. We're here at their pleasure only and mostly for diplomatic matters. Since you're not on official government business but are civilians on a strictly commercial venture, the U.S. government can't intervene with our Saudi hosts."

"I thought the State Department was supposed to help Americans in trouble," Gray said. "Well, we're Americans and we're in trouble. They're keeping us here until we pay the fine, which we can't do because we're running low on money. Can you at least lend us some money so we can take care of the fine?"

"The State Department doesn't operate that way. We can't lend you money, but if you can get your company to wire transfer the funds to us, then we can advance you some cash. But it sounds like that's not your only problem. You still have to figure out where you're going after you leave here, am I right?"

"We thought we'd fly by way of Ethiopia," I said. "We haven't talked with them yet, but we were hoping you might have some influence with the Ethiopian Embassy and could help us get an overflight clearance."

"Oh, we can't advise that," he said, shaking his head. "They're having a lot of problems in Ethiopia right now. The U.S. government definitely doesn't recommend traveling there. But the Sudan is stable, at least relatively so. I suggest you step up your efforts to obtain a clearance to fly by way of that country."

"We've been trying, but they won't answer us," I said. "Besides, the only way we can get to the Sudan is through Ethiopia or Egypt. And since our Egyptian clearance is now expired, we can't fly over that country either."

This, apparently, was not the right thing to say. The assistant ambassador rolled his eyes before continuing in an admonishing tone. "You really shouldn't be traveling in this part of the world without making some preparations, you know. These countries are very serious about airplanes flying in their airspace. You can't just fly where and when you want, and you certainly have to take into consideration that costs are considerably higher than most other places."

"Now wait a minute," Gray said. "Our company flies all over the world. We usually do have all the clearances *and* enough money. We just ran into a lot of unforeseen problems on this trip, and now we need some help."

"I don't know what else I can tell you. Maybe things will look different tomorrow, after you get a good night's sleep."

"That's another problem," I said. "We don't have any place to sleep tonight. All the hotels are booked because of the Haj. If you can't do anything for us with the Saudi government, can you help us find a place to stay?"

By now, I'm sure we sounded completely inept to this by-the-book diplomat who knew nothing about the problems of ferrying small airplanes. I knew about the problems, and I was beginning to think we sounded inept. Although he couldn't help us with the Saudi government, he was able to solve our immediate concerns by arranging for us to stay at the embassy guest quarters. They weren't much, a small room next to the laundry, but it wouldn't cost a fortune, and, more important, it was on the embassy grounds. As we settled in, Gray was still fuming over the assistant ambassador's lecturing tone and took it as a personal insult.

"It was like we were being chewed out for coming here without a clearance," he said.

"If he thinks that was dumb, it's a good thing we didn't tell him we don't have visas to be in the country."

We then began a daily routine of trudging to the airport every morning to argue with the same Arab in the traffic office. If we could just find the right person, Gray and I were certain we would be able to get them to help us get whatever clearances we needed and maybe even get the fine waived. For all our efforts, however, we were getting no further than the same Saudi Arabian clerk.

Frustrated at every attempt to resolve the situation, we would return to the embassy and be tanning ourselves at the swimming pool by early afternoon. We would plan our strategy for the next day—which didn't differ from what we had done the day before—take a quick shower and change clothes in time for happy hour at the marines' quarters. While the State Department provided us with shelter, it was the marine guards who were our source for moral support, advice in dealing with the locals, and entertainment.

"You have to go to an execution if they have one while you're still in the kingdom," a young corporal told us. He and his fellow marines had gone to one recently, and that's all they talked about.

The marines had a reputation to maintain of being tough, stoic, and fearless in the face of death. Consonant with that reputation, they attended the public executions as if it were part of their combat training. They had their own house on the embassy grounds from where they threw daily cocktail parties complete with a bottomless supply of beer and liquor. They had name brands, too, not the local bathtub gin that less well-connected Western residents had to be satisfied with.

Besides being crowded with pilgrims, Jeddah was noticeably lacking in women. Hardly any ventured out by themselves. Those few we did see—walking behind their husbands, fathers, or brothers at a respectful distance—were covered from head to toe in black robes and veils that allowed not an inch of skin to be bared.

Gray and I had been in stranger and more strict places than Saudi Arabia. But any country that stoned adulterers and cut off your hand for stealing was a place where neither of us wanted to break any laws or do anything that might appear culturally or socially unacceptable. We were careful during our forays into the city, but even that limited exposure was considered risky by the marine guards, especially because our mode of travel was public taxi.

"That's the worst way to get around," one of the guards informed us, surprised that Gray and I took our expeditions in the kingdom so lightly. "They all drive like nuts."

"They drive like that everywhere," I said.

"Yeah, but it's different here. If the taxi gets into an accident, whose fault do you think it'll be?"

Not too confidently, I answered with the presumption that whoever hit whom would be at fault. While that logic might prevail in most countries, it did not apply here.

"If the driver wasn't taking you where *you* wanted to go, he wouldn't be there in the first place," the marine explained. "So if he gets in an accident, obviously it'd be your fault."

In a twisted sort of way, it did make sense. Logic, however, had to prevail at some point. "But what're the cops going to do?"

"You're an American and the cab driver's a Saudi. Wha'dya think they're gonna do?"

"They're not going to cut our tongues off and sew them to our belly buttons, I hope," said Gray.

"If I were you, I wouldn't hang around long enough to find out. If you get into an accident, just get the hell out of there and don't look back."

Heeding our hosts' advice, Gray and I traveled light when we made our daily pilgrimage by taxi to the airport. We sat up against either door, hands resting on the handles and ready to bolt at the first sign of trouble. The ride, always through the busy city center and always in heavy traffic, seemed to get worse each day. It was especially bad at mid-morning when loudspeakers broadcast the morning prayers. Cars, trucks, buses, and all manner of conveyance stopped in mid-traffic, Arabs in flowing robes and Western business suits stopped whatever they were doing and faced east, and everyone began chanting in unison with the electronic voice.

I had heard of the Haj before. At least I think I had. I knew that pilgrims came to Mecca for some religious ceremony. I just didn't know there were so many of them or that it was that important an observance to summon so many believers. They came from Africa, the Far East, and the Pacific. The city was mixed with the many customs, languages, and cultures of distant followers of the faith, and as such, was a breeding ground for germs. While I managed to avoid catching some horrible disease heretofore unknown to Western man, Gray was not as fortunate. We had been in the kingdom five days when he woke up moaning and sneezing with all the symptoms of a full-blown cold that sent him straight to the embassy infirmary. The young nurse took one look at him and asked to see his inoculation card.

"My God!" she exclaimed, in shock unbefitting one who otherwise appeared calm and efficient. "Your cholera and typhoid shots have expired."

"Really?" said Gray, moaning again for effect and trying to sound as if he was as shocked at this revelation as she was. He was not very convincing.

"Yes, really," she said, in a stern doctor-patient tone. "What are you doing traveling in this part of the world without these shots? Especially during the Haj."

"Well, uh . . ." Gray sheepishly began, but she didn't wait for an excuse.

"These pilgrims are coming here from all over, and they're bringing malaria and all kinds of diseases with them. There's a typhoid epidemic in the Pacific where a lot of these people are from. You should at least have a typhoid shot."

"I guess I've been too busy," Gray said.

"Too busy?" she said, tersely, and emptied several needles into Gray's arm as I tried to shrink into the background. Although I hadn't looked at my shot card in a while, I suspected it was no more current than Gray's.

With Gray feeling better the following day, we settled back into our routine of airport, swimming pool, and happy hour. McCollum had extended our

Egyptian overflight clearance and wired us more money, so at least we weren't broke. However, we had decided not to pay the fine until we had someplace to fly to. In that area we were having no luck whatsoever. We couldn't get anyone at the Sudanese or the Ethiopian Embassies to even talk to us, let alone give us clearance to fly through their country. Nor were we able to get any further up the Saudi Arabian bureaucracy than the clerk in the traffic office who kept insisting he was in charge. All in all, the situation did not look promising.

The only consolation was that we were becoming famous among the embassy personnel, whose concerns usually had to do with affairs of state, protocol, and how many months before rotation back home. Gray and I gave them something new to speculate over. Everybody was interested, but nobody knew what to do with us. We hadn't met the ambassador yet, but we figured he knew about us and our problem.

Eventually, the marine guards stopped asking to see our passports and just waved us through the gate. I was beginning to feel like part of the staff, so it didn't come as a complete surprise when we were invited to the ambassador's residence for Thanksgiving dinner.

For one day, at least, we took a break from airplanes, officious clerks, and the logistical problems inherent to the Middle East. We couldn't completely forget about our problem, however, because people wanted to know if we were making progress. One of them, who had yet to hear the entire story, was the air attaché.

"So, you're the two guys in all this trouble with the Saudis," he said.

"I don't know if I'd call it trouble," I said. "It's more like we're nonpersons." He was curious about our predicament and listened as we related the difficulties we were having with the clerk at the airport.

"Sure. Everybody's in charge. If you talk to the guy who sweeps the floors, he'll tell you he's in charge. That's how things work here. Nobody wants to admit that anyone is actually in charge of them."

"If everyone's in charge, how does anything get done?" I asked.

"Oh, things get done eventually. Arabs just put a different value on time. Maybe you should leave your airplanes here and take a commercial airline back to the States. You might have better luck getting this straightened out with the embassy in Washington."

"We can't really do that," Gray said.

"Why not?"

"We can't get on an airline because we'd have to go through customs and tell them we're leaving the country."

"What's wrong with that?"

"We don't have visas for Saudi Arabia."

"You've been here a week, and you don't have visas," he said, shocked. "How in the world did you get through customs?"

"Well . . . we never actually went through customs," I admitted.

"Boy, you guys *are* in trouble," he said, and then wrote something on a piece of paper and handed it to me. "There's a sheik running everything over here. Call this guy and tell him I sent you. He's the sheik in charge of the Transportation Ministry. If anyone can help, this is the guy."

"Yes, I see," the rotund sheik said, as Gray and I told him how we came to be in Saudi Arabia but were unable to go any farther. He examined his fingernails, looked out the window, nodded his head a few times, and kept shifting his position behind the massive polished mahogany desk.

Okay, so the problems of two Americans weren't of monumental importance in his ordered existence, but at least he was hearing us out. We hadn't bothered to telephone first and chance getting an appointment next week or next month. Instead, we stopped by his office the day after Thanksgiving and told the receptionist that we *had* to see the minister. We had been working up to this moment for a week, and now here we were in front of the one guy who could either help us or not. We gave him our best pitch, and all the while he sat there only half interested with this smug Cheshire cat grin on his chubby face.

"I am sorry," he said, when Gray and I finished our story. "But I do not see where I can do anything for you."

"You're the minister of transportation. If you can't do anything, who can?" demanded Gray.

"Well, yes, I am in charge of transportation and the airports, but you did land without clearance."

"But Cairo told us we had clearance," I said. "Jeddah Approach Control didn't say anything to us. Neither did the control tower, and they cleared us to land. All we did was what they told us to do."

We were fired up now. This was our last chance, and Gray laid into him, no diplomacy, no kid gloves, just an edge to his voice that betrayed how fed up he was with this trip, the Middle East, and all bureaucracies. "And we've been arguing with that idiot at the airport every day and getting nowhere. We just want to get out of here and finish this trip. Either fine us and get it over with or help us get out of here, but we've had it."

We had the sheik's attention now. The careless grin was gone. He looked at Gray and me, no doubt shocked that we would talk to someone of his esteemed station in such a manner. He said nothing but picked up the phone and began talking in Arabic. Gray and I waited in silence. Thankfully, Gray hadn't slipped up and used the more derogative term for peoples of Arab descent favored by the marine guards, for that certainly would have sealed our fate. But still, I suspected we had gone too far.

The minister made another phone call and spoke for several minutes before turning to Gray and me, white teeth gleaming as the smile returned. Here it comes, I thought, expecting any minute for some goons to barge into the office and bounce Gray and me into the street or cart us off to the nearest prison. Instead, he told us he was waiving the fine and had just arranged with the Sudanese Embassy for them to grant us an immediate clearance to overfly their country.

Gray and I were stunned. Not only had we gotten what we wanted and more, but we were actually leaving. We thanked the sheik, who pointedly reminded us to be more diligent with clearances in the future. That evening, we went to one more happy hour at the marines' house, said our good-byes to everyone at the embassy, and left for Khartoum.

The memory of getting turned back at the border was still fresh, which caused Gray and me to wonder if it wouldn't happen again. This time I didn't think even Saudi Arabia would take us back. But no one said anything, neither in the air nor when we landed. They refueled our aircraft and let us leave for Dar es Salaam without even mentioning a clearance. It was a good thing, too, because we still didn't have a clearance number, only the sheik's word that it was okay.

Apparently, that was enough. For what Globe Aero, the Sudanese Embassy in Washington, and the U.S. State Department couldn't do, the sheik accomplished with two phone calls. The air attaché was right—all you had to do was find the right person and anything was possible. It had taken Gray and me eight days, but finally we did find that person, and through no special talent on our part other than luck.

Did Gray and I learn anything from this? Probably nothing that would make us more attentive to clearances or the borders of sovereign nations. However, we did come away with an understanding that no matter where ferrying small airplanes took us, if we got into trouble, while we might get a little help from others, we had only our wits and luck to rely on.

16

The Boom Years

The National Weather Service called the storm of 1978 the worst blizzard to hit the northeast United States since they began keeping records. People were stranded at work and on the subway. Airports closed down, and cars were abandoned on the highways to be quickly buried under the mounting snow. The National Guard was called out, and after the cleanup someone made a bundle selling T-shirts that read: I SURVIVED THE BLIZZARD OF '78.

While all this was going on, I was a thousand miles away in a hospital room in Lakeland recovering from an accident that totaled my car and left me with a concussion, a skull fracture, and something the doctors called a subdural hematoma. I spent two weeks in the hospital, the first week in a coma and the second week trying to figure out what was real and what wasn't. Every time I turned on the television something even more unbelievable was happening. It had started with the blizzard, then world heavyweight boxing champion Muhammad Ali, the self-proclaimed "Greatest," lost his title to Leon Spinks, and, finally, there was the revolution in Iran that would topple the Shah and end his two-decade reign.

I took two months off to recuperate, hoping things would settle down into some semblance of normalcy. But they didn't. Things were happening that, in recent history, were without precedent. Inflation was heating up in the United States, and interest rates were in the double digits. The price of gold was skyrocketing, and every time President Jimmy Carter made a speech, financial markets went into a tailspin. Even the local newspaper was beginning to look like a supermarket tabloid with stories about an oddball character who sneaked up on his female victims and poured spaghetti sauce on them. They called him the "Ragu Raider."

As the year progressed, it got crazier. The latest bit of absurdity was the coup d'état in the Comoros Islands. I had never even heard of the place, but now this small ex-French colony off the east coast of Africa was front-page news. What made this coup d'état different from all the others that had taken place in Africa over the past two decades was that after ousting the standing government, instead of turning the country over to the people who hired him, the commander of the mercenary army made himself the new president.

"You have to admit, even for Africa, this is weird," Holmes said, when he visited me at home to show me the newspaper article.

The way I saw it, this was further proof that the world had gone nuts. "I should've stayed in the hospital," I said to Holmes. "The year started out wacky, and it's not getting any better."

"This is just normal end-of-the-decade craziness," he said.

Whether it was the insanity that traditionally followed the calendar's ten-year cycle or the gods having fun with us mortals, there was no mistaking that this year was starting out wackier than any other. I went back to work, only to learn that aviation and our small world of ferrying was not immune. Three aircraft en route to Central America were approaching a cloud mass over Honduras when two of the pilots decided to climb on top of the clouds while the third, an ex-navy pilot named Don Walsh, descended to fly underneath and was never heard from again. At the time, we figured he must have flown into a mountain and crashed in the jungle, although the wreckage was never found. Walsh's wife even went to a fortune-teller, who told her where her husband had gone down and that the search should be concentrated in that area.

Four of Globe Aero's pilots flew to Honduras and spent several days searching the jungle by air. They found nothing, and eventually the search was called off. It took a few months, but when the rumors started coming in, they proved to be in keeping with the craziness of the year. There was talk of an insurance policy that Walsh had taken out before he disappeared. Then his wife began receiving credit-card bills with Walsh's signature on them dated *after* his disappearance. And finally, an ending so bizarre it was worthy of Alfred Hitchcock: Walsh's widow was seen at the funeral with an airline ticket to Honduras.

The belief that Walsh had faked his death and was now hiding out in some banana republic sipping margaritas was possible, though not very probable. More likely he was dead, and his body was rotting in the jungle after being picked clean by scavengers. Experiencing the death of one of our own, however, made everyone at Globe Aero, if not more careful, then at least more likely to think about the risks they were taking. As for me, coming so soon after my accident, I decided that I would be the safest pilot who ever crossed the ocean.

I began performing more thorough preflight inspections, studying the weather more carefully, and being more selective in when I took off. If an engine was going to quit, it usually happened when a major power change was made, like during or just after takeoff, a time when darkness or lousy weather can turn even a minor problem into a catastrophe.

It would therefore seem obvious that the best time to take off was when the chances for survival were greatest—when immobile objects like trees and buildings could be seen and, hopefully, avoided. One might think this would lessen, if not eliminate entirely, the practice of taking off at night or in fog. It did for me, but only for a short while. After several months, during which I experienced no engine failures or problems of any kind, and since I was more concerned with the time of day I wanted to arrive at my destination than with the possibility that something might go wrong, I lapsed into my old habit of taking off at night. Whether because of luck or because the Trim God had decided to look favorably on me, the only time I had an engine failure on takeoff was during daylight.

It was early spring when Gray and I left Gander en route for Amsterdam in two identical Beech Bonanzas. Climbing out under a crystal-clear, midday sky, the Newfoundland landscape unraveled below us in a patchwork of green and brown. The gray-blue waters of Lake Gander a half mile to the south had shed its last trace of winter ice and was as smooth as glass. Even with the weight of the extra fuel, the single-engine Bonanza climbed at a respectable eight hundred feet per minute.

I passed through twenty-five hundred feet and looked forward to an easy crossing when I heard the one sound I hoped I would never hear. The engine sputtered, then surged . . . once . . . twice, and finally quit. I quickly turned on the fuel boost pump, but nothing happened. I switched fuel tanks and checked the vapor return valve. I tried restarting the engine, but the propeller only spun feebly in response. Everything looked normal, but no matter what I did, I couldn't get the engine started again. The radio was busy with aircraft taking off and landing, but it became deathly quiet when I transmitted the single word: "Mayday."

After telling the controller the nature of the emergency, I added with resignation not fully convinced of the outcome, "I'm turning back toward the airport, but I don't think I'll be able to make it."

There was a brief pause before the controller advised me that rescue crews had been alerted. This, however, didn't have the impact it probably should have because I was too busy trying to find a place to land while keeping the airplane under control. I lowered the aircraft's nose to keep from stalling, knowing that with the weight of the extra fuel I would come screaming down faster than normal. But I *had* to maintain flying speed. A controlled crash landing under any circumstances was survivable, whereas letting the airplane stall and spin out of control, I knew, would be fatal.

The end of the runway was only about four miles in front of me, but as fast as I was losing altitude, I figured I would land a mile short. I began looking for some place to set the airplane down. Trees were everywhere. The lake was too far, and there were no roads that I could try to squeeze onto. The airplane was dropping rapidly when I saw the clearing. It was small and strewn with boulders, but at least it was flat. With luck I might just be able to reach it.

At three hundred feet the ground began taking shape. It looked hard and not as level as it did when I first spotted it. I knew then that I would never get out of this without causing *some* damage; hopefully, I could keep it to a minimum. Distance to the clearing was closing quickly, as I gauged speed and altitude, trying to control each so that I would not run out of one before the other. I was about one hundred feet above the ground when I lowered the landing gear, hoping to make as normal a landing as possible.

As soon as I felt the wheels touch down I knew I had made a mistake. What had at first looked like hard, solid earth turned out to be soft and muddy. The airplane sank up to the wings, snapping the landing gear off as momentum and mass propelled the fuselage forward, skidding over boulders and bushes before coming to an abrupt stop. My upper body, however, kept moving forward, until my head slammed into the windshield.

It took a minute or so before I was able to think clearly, and then my first thought was fire. Quickly unbuckling my seatbelt, I crawled over the fuel tank and out onto the wing and stumbled onto the ground. I could hardly believe what I had walked away from. One wing was bent upward, and the engine compartment was twisted. Shorn of its landing gear, the airplane crouched on the ground like a mortally wounded eagle.

A Canadian Rescue helicopter touched down about twenty-five feet away, and two paramedics came running over. "Are you all right?" one of them asked.

"Uh . . . don't ask me any hard questions right now," was all I could stammer.

The paramedics helped me onto the helicopter and flew to Gander Hospital, where I was rushed to the emergency room, dazed and probably in shock. Physically, I felt okay, so I couldn't understand why everyone was making such a fuss.

"You came very close to losing an eye," the doctor said, as he held up a mirror so that I could see the damage.

The face looking back at me was mine, all right, but it did not look good. There were scratches and bruises everywhere, a large bump that was already turning purple, and a deep gash on my forehead ending just above my right eye that was big enough that I could stick my finger in. In my dazed condition, the seriousness of the cut and how close I came to losing an eye didn't register. I lay there half-bored as the doctor stitched me back together and put a bandage over my eye.

"We're going to keep you here overnight so we can take a look at that cut in the morning," the doctor said, and handed me off to the orderly. I was wheeled into a room and helped into bed. No sooner had he left, than Gray and Briggs walked in, sat down beside my bed, and stared at me for what seemed a long time.

"Well, how do you feel?" Gray asked. "You *look* like hell."

"I've had better days. But you should see the airplane."

"I did. I watched the whole thing from the air. I told the tower, 'that's my buddy in that airplane.' So they let me circle until you got on the ground."

"Pretty good landing, huh?" I said, trying to keep a sense of humor.

"Oh, the landing wasn't bad, but you only rolled about twenty feet. Then I saw you get out of the airplane and fall off the wing, and I got scared. But you got right up," Gray said, chuckling, "so I figured you were okay. What happened, anyway?"

"Beats the hell out of me. The engine just . . . quit. It seemed like it wasn't getting any fuel."

"Did you have all the valves in the right position?" Briggs asked. He landed just after my crash and hurried over to the hospital.

I couldn't remember *what* position they were in, and all I could say was, "I hope so."

"That's the first thing I checked," said Gray. "I landed right after you did and ran out to the airplane to make sure everything was the way it was supposed to be. All the valves and switches looked okay, but you could've picked a better place to land. I was up to my ankles in mud."

"How's the airplane?" Briggs asked. "I saw it from the air, but the tower wouldn't let me get too close."

"It's totaled all right. I'm surprised it didn't blow up, because both ferry tanks buckled and slid forward," Gray said, as he shook his head in disbelief. In reference to my obvious forward trajectory after the airplane came to a stop, he added, "You probably weren't wearing your shoulder harness, were you?"

"Probably not," I said.

"What do you mean, probably not?" Briggs demanded. "You're always supposed to be wearing your shoulder harness."

"You probably would've gotten away with no bruises at all if you had it strapped on, you know," Gray said.

"Yeah, I know," I agreed, now feeling guilty.

"Look at the bright side," said Briggs. "All you need is an eyepatch, and you can get a job as an extra in a pirate movie."

Laughing over Briggs's suggestion, I must have strained some muscles in my nose, because blood started gushing out. A few minutes later, a nurse came in and gave me a towel to stop the bleeding, adding a warning to hold my head up. Meanwhile, Briggs, ever conscious about costs while on a trip, was assessing the sparsely furnished room.

"What're they charging you for this place?" he asked. "It probably costs a fortune."

"This is Canada," I said nasally, looking up at the ceiling. "Socialized medicine. The doctor was free, and the room's only twelve dollars a night."

"That's all?" he exclaimed. "I should start staying here instead of the Albatross."

"The only catch is, you have to crash."

"Which is something I really wish you hadn't done," said Gray.

"To be honest, I had very little to do with it."

"Yeah, but I still have to make the crossing. Both our airplanes were going to the same place. They came off the production line together, so if something was wrong with your airplane, chances are I'll have the same trouble with mine."

Realizing Gray's concern, Briggs offered his usual response for crossing the ocean when all looked bleak and he wasn't the one with the problem. "Don't worry. You'll probably make it."

Gray decided to leave for Amsterdam before he could talk himself out of it. After he and Briggs left, John Newhook, one of the new weather briefers, stopped in to see how I was doing. Everyone wanted to visit and see if I was okay, but all I wanted to do was sleep. Not five minutes after Newhook left, a Catholic priest came into the room. Now I was wide awake. I was also scared to death.

"Hello, my son," the priest said. "Are you comfortable?"

I eyed him suspiciously. "Yes," I lied. I really wasn't comfortable. Why was a priest visiting me? I wondered. I had seen enough movies to figure that out. My wounds were more serious than I thought, and the doctors, and probably even Gray and Briggs, were keeping the truth from me. But I knew better.

"I know why you're here," I said, accusingly, as if I had just caught him pinching the collection box.

"What do you mean?" he asked, furrowing his brow inquisitively.

"You know," I said.

"No, but maybe you could tell me"

I had him now. "I'm going to die," I said, the emotion gone from my voice. "And you're here to give me last rites."

"Oh, no! Is that what you think?" The priest's look of surprise turned into a grin as he tried to reassure me, "I'm here to visit *all* the patients."

Sure, I thought, but said nothing. What was there to say?

"I'm sure you're going to be all right. What happened, anyway?" he asked. "They told me you were in a plane crash."

"My engine quit," I said, leery of this line of questioning whose purpose seemed only to take my mind off the inevitable. "I guess I must've angered the Trim God."

"Did you say Trim God?"

"Yeah."

"And who might this . . . uh . . . Trim God be?" the priest asked, hesitant at my incursion into his area of expertise.

"He's the god who watches over ferry pilots."

"There is only one God, my son," the priest informed me with righteous confidence.

"True. But running the universe is a pretty big job. He's probably got a lot of lesser gods that help out."

"But that's paganism!"

"Then I'm a pagan," I said.

I could tell that he felt uncomfortable, because he kept fidgeting and trying to think of something to say. If he felt uncomfortable, how did he think I felt? I was the one in a plane crash, and I was the one who was going to die. Although he did his best to reassure me that his visit was in no way meant to prepare me for the final journey, I didn't trust him. He suggested that I get some rest and left shortly thereafter.

Rest was the last thing I wanted. Convinced that if I fell asleep I wouldn't wake up, now I *wanted* to stay awake. It was no use; I was just too tired. I gave up after a while and eventually dropped off to sleep.

I was never more relieved to wake up the following morning to the sound of doctors and nurses scurrying about their duties, and not the sounds of a celestial choir or something far worse. I was alive—a condition I had every intention of prolonging. So far this year I had been in a car accident and now an airplane. There was only one thing left—boats. I decided right then that I wasn't going anywhere near the water. That meant no more water skiing in Holmes's boat. But I had been backing away from that sport anyway ever since I learned that all the lakes we skied in were full of alligators.

Satisfied that the gash over my eye would heal properly, the doctor gave me a final warning not to scratch it and discharged me from the hospital. Before I left, I went into the dispensary and bought a black eyepatch. As I flew home on the airliner, I'm sure I must have been a sight to the little kids who pointed and stared open-mouthed. All I needed was a peg leg and a parrot perched on my shoulder.

Whatever the problem was with my airplane, it remained an isolated one. Gray made it safely across the Atlantic and landed in Amsterdam without incident. But he didn't stay with Globe Aero much longer. Like many of the other ferry pilots, Gray's goal was to fly for the airlines. He had been on several interviews and finally landed a job with Allegheny Airlines.

"You'll get bored," I said, when he called to tell me the news.

"Yeah, but I have to do it," Gray said. "If I stay in ferrying much longer I'm going to get spoiled, and I won't want to do anything else. Besides, these distributors can't keep buying all these airplanes indefinitely. It may be busy now, but it has to slow down sometime. I think what we're seeing is the last hurrah in the ferrying business."

Gray was right on both counts. We *were* ferrying a lot of airplanes, more than we had ever ferried in previous years, and if you stayed in this business too long you got so hooked that even if you had the chance you wouldn't want to get out. I think I had already reached that point. I had been ferrying airplanes for only five years—not a long time, but long enough that I didn't think I could adapt to flying a schedule that repeated itself day after day. If this bubble were going to burst some day, as many of us suspected, then we would worry about it at that time. The goal now was to enjoy it while it lasted. Except that it was hard to enjoy when so many of my friends were quitting or dying.

No one expected it to happen. Not to Joe Wolf—he was indestructible. He flew with Walt Moody back in the early days when Globe Aero was just getting off the ground, and the only other pilots were Phil Waldman and

Bob Campbell and ferrying small airplanes over the ocean was considered so risky that it couldn't be done, let alone that someone would do it several times a month. Wolf died when his airplane wrapped around a tree stump after his engine quit one night and he made a forced landing into what he hoped was an open field.

After all the trips he made and the chances he took over the ocean, to end his life over land on a simple flight from Vero Beach to Lock Haven was an irony that none of us understood. Didn't we all take the same risks and push ourselves just as hard? Didn't I, didn't Kerby, Gray, Holmes, Briggs, and every other ferry pilot sleep over the ocean when we should've been awake more often than we cared to admit?

Wolf's death was a reminder of just how vulnerable we were to the designs of the Trim God. If it happened to him, it could happen to any of us, no matter what we did. Wolf, on the other hand, never seemed to heed the warnings. No close brush with death or slump in business kept him on the ground; for buried somewhere in his Rolodex was a customer who needed an airplane ferried. It was ironic that Wolf, who would fly any airplane to any country at any time would miss the busiest years ferrying had ever known.

Deregulation of the airlines was less than two years old, and the first major expansion the industry had seen since the late 1960s was beginning. Airlines were hiring again, and that precipitated a chain reaction throughout the aviation world. Flight schools were materializing overnight, and every one of them—civilian, airline, and military—needed new airplanes to train this army of pilot hopefuls. As these students progressed to private and commercial pilots, they demanded bigger, faster, and more sophisticated airplanes. Good capitalists that they were, manufacturers rose to meet the demand.

During the years I had been flying for Globe Aero, we delivered between two and three hundred airplanes each year. Now we were closing in on four hundred. Airplanes were piling up on our parking ramp, yet Piper kept calling with still more aircraft waiting to be ferried overseas. The business that Walt Moody started with just a few European and Australian customers now included aircraft dealers and distributors, corporations, airlines, flight schools, the U.S. and foreign governments, and militaries throughout the free world.

"I need some warm bodies," Waldman would say, and then start hiring squadrons of pilots. It didn't matter if they had experience or not. If they could walk into the hangar under their own power, Waldman put them to work following one of the senior pilots. Now, after five short years, the other pilots and I who had started flying for Globe Aero in the mid-1970s—like Campbell, Wheeland, Mantzoros, and Wolf—were the old timers. And like them, we looked at this new group of pilots with the same jaundiced eye I'm sure we had provoked.

"Were we ever as green as these guys?" we said to each other, while cursing their inability to follow even the simplest instructions. In all likelihood we made the same mistakes ourselves; however, now all we could do was shake our heads and wonder, "Where did Phil ever find this guy?"

With twenty-three pilots, each flying four to five trips a month, it would've been difficult at any given time to find an ocean without a Globe Aero pilot crossing it. Not just in groups of two or three, whenever we went anywhere these days there might be up to a dozen aircraft flying in one huge gaggle, crisscrossing the planet from Tokyo and Stockholm in the northern hemisphere to Perth and Capetown in the southern hemisphere.

As always, most of our trips were over the Atlantic, an ocean that had rarely seen a shortage of airplanes. It was especially crowded now, with airplanes bound for Europe, Africa, and the oil-rich sheikdoms of the Middle East. And almost every one of them landed in Gander. When storms over the ocean made it foolhardy to even attempt a crossing, airplanes started piling up on the airport parking ramp, each day producing yet another, whose pilot had to wait out the weather, sometimes for several days. With so many new pilots flying to Gander, it was often difficult to keep track of who flew for whom. It was not uncommon to be sitting at one of the hotel bars and strike up a conversation with the stranger sitting on the neighboring barstool, only to learn that he too worked for Globe Aero.

Gander was one of the few places in the world where ferry pilots gathered in such large numbers. I could just about always count on running into someone I hadn't seen in months. The sole exception was Andy Knox. It seemed as if every time I flew to Gander and checked into the hotel, Knox was already there. He was beginning to look like a permanent fixture. Knox was in his sixties when he started ferrying airplanes for Globe Aero three years earlier. After taking a year off, he went to work for Wolf's company, where the first airplane he ferried had the German registration letters D-EATH.

A person didn't have to be superstitious to feel uneasy about flying a single-engine airplane over the ocean with the Grim Reaper's name painted on the side in bold twelve-inch letters. Two other pilots had already turned down the trip before it came around to Knox. Less concerned with ominous warnings than with the money he would make, Knox ferried the airplane and became an instant celebrity with air traffic controllers when he called in over the radio and gave his call sign phonetically.

After routinely repeating the letters, the controllers realized what they had just spelled and usually made some quick-witted comment. Knox then told them he was taking the airplane over the ocean, which invariably prompted any airline pilot on the frequency to question his sanity at providing such a tempting target to fate. After several such calls, Knox dropped the phonetic spelling and began his transmissions with the more direct, "This is Death calling."

Gray-haired, unhurried, and with a cigarette always nearby, Knox braved the long hours in cramped cockpits with the same stamina as men half his age. Where he got this energy is a mystery, because the only exercise I ever saw him get was from bending his elbow with a scotch and water when he held court in the lounge of the Sinbad Hotel.

Sinbad's was Gander's newest hotel—conveniently located in the center of town and staffed by employees who exhibited an aptitude for meteorology. As soon as it started to snow, they set aside several rooms for ferry pilots. Knox was at his usual table with several of the new pilots when I walked into the hotel's lounge. Kenn Dawson and Geoff Tyler, two ex-flight instructors from Vero Beach, were there, along with Rich Houghton, who was with Don Walsh when he disappeared inside the cloud, Tom Willett, an ex-air force navigator, and Lisa Kilbourne, Globe Aero's first and only female pilot.

Based on the number of empty beer bottles and whiskey glasses, it was obvious that weather over the ocean did not look promising. Tonight everyone would do what they usually did when more than two or three pilots were present—get drunk. Eventually, if past experience was any indication, the inebriated group will wander over to the Flyers Club to catch the latest stripper. Prior to that, however, there was gossip to be exchanged and analyzed, at which point insight was drawn and conclusions reached. The topic on this particular evening had to do with the budding romance of Stan Finegold, a nineteen-year-old pilot whose goal was to fly every ocean before his twenty-first birthday.

"This is bad," Tyler lamented. "Finegold's just a kid. One of us has to pick him up at the airport when he comes off a trip because he's too young to rent a car. And now he's fallen in with fast women." Older and presumably wiser, Tyler felt it his responsibility to watch over the young pilot, lest his naivete lead him astray.

"Fast women? In Gander? I wish someone would tell me where they're all hiding," said Willett, who, since coming to work for Globe Aero, seemed determined to unseat Flournoy's reputation as the resident lady's man.

"He's just dating the girl," said Kilbourne, who held her own in this mostly-male profession.

"That's how it starts," said Tyler. "Pretty soon he'll be coming up to Gander all the time. Then he'll move in with her, get married, and bang—before you know it, he's got two kids keeping him up all night changing diapers."

"Maybe he's just working up to that hundred dollar prize you offered to the first pilot who goes to bed with an Aeroflot stewardess," suggested Houghton.

"Not Finegold. These Russian stewardesses would have him for breakfast," argued Tyler. "I tell ya, we gotta do something before it's too late."

"Moriarty's already on top of it," said Dawson. "We were right in this bar one night having the same conversation, when he says, 'Who is this girl, anyway? What do we know about her? We gotta go check her out—make sure she's right for him.' "

"That sounds like Moriarty," said Knox. "So did he?"

Dawson gave a quick laugh. "Yeah. The next day he went to the store where she works and started asking her all these questions—like he was interviewing her for a job."

"Well, we don't want him to end up like Holmes," said Houghton.

Holmes was often the example brought up when the logistics of a pilot's trip was determined less by weather or geography than by the location of a particular member of the opposite sex. The trouble he had gotten into in Honolulu with his girlfriend's father was well-known, as was the reckless way he used to fly his trips just so they could spend more time together. Holmes still rushed his trips, only now he was in a hurry to get back home because he eventually married the girl and brought her to Lakeland.

"Where *is* Holmes, anyway?" I asked. "I thought he was on his way up here?"

"He's already been here and gone," said Willett. "He took one look at the ocean and said, 'Well, see you guys in Lakeland,' and left for Reykjavik. Course, he was in a Cheyenne, so he could get above the weather, but still, I don't know how he does it. I've never seen him take more than three days for an Atlantic trip."

"I was with him in the Pacific once; you should see him out there," said Kilbourne. "He shows up at the airplane in Honolulu with two bags of groceries and then goes straight to Sydney, stopping only long enough to refuel. He doesn't waste a lot of time sleeping, does he?"

"Holmes was born on Krypton," I said.

"He's gonna wind up like the Dead Guys," said Tyler.

"The who?" I asked.

"Didn't you see those two guys back at Lakeland?" Tyler asked. "They were getting their airplane tanked at Globe, a single-engine Mooney that looked like it'd been used as a chicken coop for the last few years." He shook his head and laughed. "I wouldn't even fly it from Lakeland to Orlando."

"Neither of them had ever made a crossing either," said Kilbourne. "One of them had just gotten his instrument rating, and the other one had a grand total of about six hundred hours."

"And they were going across the Atlantic? This time of year?" Knox asked, amazed that anyone would attempt such a flight without *some* training.

"Not only that, but they were planning to take the northern route through Greenland and Iceland," said Willett.

"What're they, nuts?" exclaimed Knox, setting down his empty glass and signaling the waitress for a refill. "This is no time to be flying to Greenland, especially in a single-engine. I don't care how many crossings you've made. Wait 'til they have to scud run down the fjord a hundred feet off the water."

"Or they have to land at Reykjavik in fog with a fifty-knot crosswind blowing ninety degrees to the runway," I added. "That'll make believers out of them."

Knox and I were not merely speculating about arbitrary disasters that could befall a pilot flying the North Atlantic in winter, but citing actual events that had happened to each of us. Yet, we refrained from boasting because no matter how dangerous a crossing we might have had, another pilot had a story that was even more outrageous.

"Kerby tried to tell them, tried to talk them out of it, but they wouldn't listen," Tyler continued. "These guys were dead set on going, and there wasn't anything anybody could say to talk them out of it. So after a while, we just started calling them the 'Dead Guys.' "

"After the revolution, they're not going to let anybody over the ocean by themselves until they've got a hundred crossings," I said, with the confidence of having one hundred and thirty crossings in my logbook.

"And what revolution is that?" asked Kilbourne.

"The world revolution," I said.

"Interesting theory," speculated Dawson. "But is there going to be a reason for this revolution, or is it just going to happen on Karl Marx's birthday?"

"Of course. The usual reasons," I said. "Class warfare—the haves against the have-nots. The rich against the poor. The bourgeoisie against the working class."

"Republicans against Democrats," suggested Tyler.

"Even the French?" a skeptical Willett wondered.

"Everybody. Because this is gonna be the big one," I continued. "There's going to be looting, fires, bombings, chaos, panic. Financial markets will plummet, and they'll be mugging old ladies in the streets!"

"Hey, I like this. Whose side are we gonna be on?" said Tyler.

Any mention of revolution eventually brought the conversation around to the Russians and Aeroflot, the Soviet airline that stopped in Gander on its Moscow-Havana flight. Unlike other flight crews that overnighted in Gander, the Soviets didn't stay at any of the hotels or eat in any of the restaurants. Before glasnost, when the Russians were our bitter enemies and their self-imposed rules of the cold war precluded association with any foreigners, Aeroflot crew members stayed in their own apartment house and ventured out only when absolutely necessary. This isolation so incensed Tyler, that—following the example of President Carter in his attempts at diplomacy with the Soviet Union—he wanted to bring a bottle of vodka to the Russian crew house to do our part for détente. We never did learn how successful such a gesture would be in improving diplomatic relations—at least as far as the aviation community in Gander was concerned—because we never quite reached the state of inebriation necessary to actually put this talk into action.

Neither did we find out if the Dead Guys ever made it to Europe—although we knew that the odds were not in their favor. It was pretty much an accepted fact that a certain number of ocean crossings would result in a ditching. That was, after all, proven by the laws of probability. Although an experienced pilot might be able improve those odds by virtue of his or her flying skills—not to mention their knowledge of weather and navigation—if you had to ditch, survival depended more on luck than anything else.

Of all the pilots who flew the oceans, whether amateur or professional, none was luckier than Peter Goldstern, a New Zealander who went down in the North Atlantic in the dead of winter and survived. Given that he ditched at night and in mid-ocean was remarkable enough, but what made Goldstern's tale even more incredible was what happened *after* he was rescued. While the efforts of air traffic control and Canadian search-and-rescue crews deserve much of the credit for Goldstern's survival, also in his favor was that instead of a life raft he had a new type of immersion suit with him. Looking like an astronaut with a spacesuit several sizes too large, the suit kept him dry and afloat for three hours until a Royal Canadian Air Force rescue aircraft homed in on his emergency transmitter and dropped a twelve-man life raft.

Six hours later—cold, hungry, and weak—he was picked up by the *Georgy Ushakov*, a Soviet weather-observation ship known to the briefers in the Gander met office as Ocean Station Charlie. Since the vessel stayed a month at a time on station and had arrived only three weeks previously, Goldstern had to spend the next eight days as a guest on the ship.

It was New Years Eve when the *Georgy Ushakov* began the return journey to Russia with the intent of dropping their passenger off in Genoa, Italy. When they got within twenty-five miles of Genoa, the Italians telexed the ship that permission to enter Italian territorial waters had been cancelled—apparently in connection with the Soviet's problems in Afghanistan.

Goldstern sent messages to the New Zealand, British, and U.S. Embassies, as well as the president of Italy, requesting help—none of whom bothered to reply. Several days and many phone calls later, with the help of a friend in the Argentine Embassy in Rome, the British consul—authorized by the New Zealand Embassy who was guaranteed payment by the Argentine Embassy—chartered a tugboat to transport the pilot to shore. Because his passport, along with the rest of his belongings, was at the bottom of the ocean, Goldstern had yet another round of bureaucracy to contend with before he was finally able to bring his twenty-two day ordeal to a close.

News of Goldstern's survival was inspiring. It meant death did not have to be a natural consequence of ditching. All we had to do to stay alive was keep our heads, take a few precautions, and, of paramount importance, start looking through the latest marine-equipment catalogs so we, too, could order those new immersion suits.

No matter how many pilots were rescued from certain death or how many successful crossings we made, there was one inescapable fact we lived with constantly: flying a small airplane in the subzero temperatures of the North Atlantic was asking for trouble. If ice didn't get us, there was always a risk the engine oil would congeal and become too thick to lubricate the engine. Even with our limited knowledge of fluid mechanics, we knew that an engine with no oil to lubricate its moving parts would not work long.

Thankfully, this type of problem was rare. Most of the aircraft we ferried were Pipers with reliable Lycoming engines so sturdily built that they were almost indestructible. More common than engine failures was the problem we had with heat—either we had too much or not enough. Most single-engine aircraft have what we call a muff heater, which is simply air warmed by the exhaust manifold and routed into the aircraft's cabin. No matter how cold it was outside the airplane, the inside stayed warm and put out enough heat to cook a can of soup if placed directly on the heater vent. It was a foolproof system, considering that it worked as long as the engine did. And, although engine failure would stop the flow of warm air into the cockpit, heat would not be of paramount importance.

A basic tenet of aviation is that the bigger the airplane, the more complex the systems, and the more complex the systems, the more susceptible they are to failure. This was especially true with systems designed to produce heat. Instead of a reliable muff heater, most of the twin-engine airplanes were equipped with a small, gas-driven motor that put out enough heat to combat even the coldest temperatures. The only problem was that they didn't always work. On the Navajos the heaters were notorious for failing at the worst possible time.

Too often I had been in a Navajo at seventeen thousand feet, cruising along with a hundred-knot tailwind when cold air started coming out of the heater vent. More often than not, the cause of the failure was because the heater had been running continuously to keep up with the cold and had popped the circuit breaker. While this might seem like a minor problem requiring the pilot to simply reset the failed circuit breaker, it was not possible to do during flight because the circuit breaker was on the heater motor itself. The only way to get to it was from outside the airplane.

This never failed to elicit queries from my nonflying friends as to why I continued to ferry these aircraft if the heaters kept quitting. Responding in my best devil-may-care manner, I tell them that the heaters only quit when it gets real cold, and it doesn't get that cold that often.

But it did get that cold when I was bringing an airplane from Europe back to the United States. We didn't ferry many airplanes in that direction, and the few we did were usually second-hand. This time, however, I was flying a new Britten Norman Islander. The Islander was a rugged, twin-engine, commuter aircraft with a long narrow fuselage capable of carrying ten passengers or eight fifty-five-gallon drums of fuel at a speed of 125 knots. It was an excellent airplane for what its name suggested—island hopping in the Caribbean or the South Pacific, where flights were never longer than a few hours and the temperature rarely dropped below seventy degrees. The heater was the same type as the Navajo had, but because the Islander was built by a different manufacturer with different wiring and system design, I figured it should work okay.

I picked up the airplane at the factory on the Isle of Wight in southern England and flew to Glasgow to begin the wait for the right weather conditions. Because of prevailing westerly winds, even with extra fuel, most of the aircraft we ferried westbound didn't have the range to fly straight across the ocean. Instead, we flew the northern route over Greenland and Iceland.

The North Atlantic in winter is a cold, damp, and miserable place, even on the Gander-Shannon route. It got worse the farther north one flew. There was Greenland, with its unpredictable weather and treacherous terrain. Iceland has its deceptive winter wonderland lure that could change overnight into a frigid inferno of heavy snow, gale-force winds, and fog so thick that not even light from the snow-covered city could escape. Storms pushed their way across the North Atlantic one after another. They swung down from northern Canada or came up from the Mid Atlantic states and were carried north with the high altitude jet stream. Just about every one of them passed over or near Iceland, and as one moved away, another approached. It meant we had to time each leg so that we flew in between the storms, which often took several days for conditions to be just right.

Expecting my wait in Glasgow to be lengthy, I was surprised when, after only one day, the low-pressure area that had been battering Iceland moved to the east as another low approached from the south. This second storm was far enough behind the first storm that it left a corridor of clear air between Glasgow and Reykjavik.

By flying close to the northeast side of the approaching low, I was able to stay in warmer air coming up from the south and also get a good push from the winds. At sixteen thousand feet, the winds, although not as strong as forecast, still carried me to the Faeroe Islands in just over two hours. The islands—a precipitous convulsion of volcanic leftovers spawned in the Norwegian Sea eons ago between Iceland to the northwest and Scotland to the southeast—were the midpoint of my leg and all but invisible in the dark ocean.

Strange things happen this far north, where compasses often went crazy because of the vast distances between the geographic and magnetic poles. And the Aurora Borealis cover the heavens like a giant curtain with vertical folds rippling as if in a light summer breeze. Columns of light shimmering from red to blue to yellow and to green danced so close to my airplane, I felt I could almost reach out and touch them. They grew larger, brighter, and more erratic the farther north I flew and slowly disappeared as I began the descent to Reykjavik.

I slept late into the morning, woke up in time to catch the lunchtime fashion show at the Hotel Loftleider, and then went next door to the flight services office, where Sveinn Björnnson tried to talk me into breaking up my flight to Goose Bay by stopping at Sondrestrom.

"What's the matter with Narsarsuaq?" I asked.

“The weather is not good, I am afraid. Low clouds down the fjord. But it is better to the north. The Labrador high is over northern Greenland, and so Sondrestrom will be clear.”

“And probably cold, too.”

“Ja. It is that time of year. They are reporting minus eighty degrees Centigrade,” he said, casually, as if this were something normal.

“With temperatures like that, I don’t want to go anywhere on Greenland; especially inside the arctic circle,” I said. “I’ll overfly Prinz Christian, but if everything looks good, I’m not stopping until I get to Goose.”

“But you will have headwinds. Twenty knots I think. All the way across.”

One of the unwritten laws of aviation, a law any meteorologist fresh out of school would probably dispute, was that headwinds, when encountered in flight, were usually stronger than forecast and tailwinds were weaker. The twenty-knot headwind Björnnson had warned me about turned out to be closer to forty knots. Because I was flying directly into the wind, my speed over the ground was down to eighty-five knots. At that rate, I figured it would take more than fifteen hours to reach Goose Bay. This would not have been bad if it weren’t for the cold.

The high-pressure area over Greenland that brought clear skies to much of the hemisphere also brought a chunk of cold Siberian air that turned the northern Atlantic into one big deep freeze. Every station in Greenland and northern Canada was reporting temperatures twenty degrees colder than normal for this time of year. The temperature outside the airplane had been dropping steadily for the past hour and was now at thirty-two degrees below freezing. The trouble was, it didn’t feel much warmer inside the airplane. I knew the heater was working because I felt warm air coming from the vent. It just wasn’t putting out enough warm air to keep up with the cold.

Rummaging through my suitcase for warm clothes, I put on another sweater and wool socks, and wrapped myself in the space blanket I always took with me. It helped some, and for the next several hours I managed to keep from freezing, but then the sun went down and the temperature dropped a few more degrees. I was over Prinz Christian and still had another seven hours to fly to Goose Bay. As cold as it was, I didn’t think I could last another hour, never mind seven. I had to start looking for someplace to land before I froze to death.

Narsarsuaq was only sixty miles north and would've been perfect, if not for the clouds. If it was light out, I might have attempted to fly down the fjord, but not in the dark and especially not in the half-numb condition I was in. Kulusuk and Godthab were the only other airports in southern Greenland, but their weather was the same as Narsarsuaq. As much as I hated to admit it, if I wanted to land on Greenland, Sondrestrom was the only place. But it was five hundred miles away and would take almost as long to fly there as it would to Goose Bay. And for what? To freeze as soon as I stepped out of the airplane? Presuming I made it that far. I couldn't return to Reykjavik, because the low had moved in and now their weather was deteriorating. There was no way out of it; I had to continue to Goose Bay.

Thirty minutes later my fingers and toes were beginning to feel brittle, and a simple task like flipping a switch was an awkward and painful ordeal. I had to find warmer air, but the only way to do that was to descend. I was at ten thousand feet now. Even if I dropped down to just above the water and the temperature rose at the standard rate, the warmest I could expect would be fourteen degrees below freezing. While that would be an improvement, there was something else preventing me from descending. A thousand feet beneath me was a solid layer of clouds. Earlier, when it was still light out, the layer looked thick and laden with moisture that would freeze as soon as it came in contact with my airplane. Now, in the dark moonless sky, everything was black and murky, no longer a crystalline blanket, but no less forbidding. I would never get through those clouds without picking up ice. With the freezing level on the surface, there was no way to get rid of it.

It would be suicide, but I was too cold to care. More than once I thought of just pushing the nose straight down and getting it over with quickly rather than suffer another minute. Cold as I was, I knew that I had to do something before that idea outweighed all others. I began a descent, telling myself that I was just going to skim along the tops of the stratus. But as I got lower, so did the clouds. I leveled off just above the tops at seven thousand feet. The temperature gauge read twenty-six degrees below freezing—not what anyone would call balmy, but after the last several hours, even with this minor six degree rise, I felt warmer.

When I began the descent for Goose Bay five hours later, I was so numb from the cold that I could barely feel the rudder pedals under my feet. I kept the engine at full power as I came screaming down to the runway, only slowing just before landing. I stayed over in Goose Bay that night, but in my mind I was already back home buying long johns, bulky turtlenecks, and an extra-thick parka.

Later that evening, as I endeavored to further warm my insides with one brandy after another, my thoughts changed from warm clothes to cold revenge. I didn't know how it would happen, but someday I would have a representative of the heater manufacturer as a passenger and would be able to demonstrate—by cold, miserable example—the reliability of his product. In the meantime I tried to fix the problem myself. Whenever I got an airplane with an electric heater, I rewired it so that I could reset the circuit breaker from inside the cabin. While it tested okay on the ground, I never could get it to work in flight and eventually quit trying. There were too many other things to worry about. With the arrival of the new decade came more pilots and small airplanes over the oceans, along with an increasing number of mishaps and tragedies.

17

Farewell from the Hereafter

There was no reason Rich Houghton wasn't blown to pieces over the Sahara Desert. After all, as was so often the case when we flew in Africa, he was not where he was supposed to be. Houghton was flying from Tenerife to Abidjan on a route that came uncomfortably close to the Algerian border. Furthermore, it was during daytime instead of at night when he would've been under the cover of darkness. Maybe then he wouldn't have been awakened from his nap by the sound of a loud roar.

Thinking an engine had blown up, Houghton scanned the gauges looking for anything out of the ordinary. Everything appeared normal, so he figured he must have been dreaming. He was about to resume his restful position sprawled across the pilot and copilot seats when he heard it again. This time he saw what was causing the noise—a Mirage fighter jet crossing in front of him so close he could make out the Algerian flag on its tail.

The Mirage fired another cannon burst, maneuvered alongside Houghton's airplane, its landing gear and flaps lowered, and waggled its wings. Unsure of the meaning of that particular maneuver and not wanting to respond in a manner that might be interpreted as aggressive, Houghton checked his

International Flight Guide and read: "Indication of wings waggling means you have been intercepted. Follow me!"

Houghton waggled his wings in response and waved to the pilot in acknowledgment. The Mirage made a one-hundred-eighty-degree turn and began descending to an airport Houghton had noticed a few miles back. Thinking this would be his only chance to escape, Houghton pushed the throttles forward and headed straight for the border, and what he hoped was safety. Meanwhile, the Mirage pilot, noticing Houghton wasn't following, turned around and caught up with him just before he reached the border. Again, the Mirage circled Houghton's airplane in an attempt to get him to turn back to the airport and land.

Houghton kept the airplane pointed toward the border. It was time to put the ball in the Mirage pilot's court. He would either have to shoot him down or let him go, because Houghton had made his decision—he was *not* going back. The Mirage kept circling until Houghton crossed the border, then made a steep, climbing left turn and flew off in the opposite direction. Houghton looked in his *Flight Guide* for the meaning: "Okay, you're free to go."

Houghton had been lucky, because African countries were no longer remaining passive with airplanes that wandered into their airspace. Armed with all the weapons of modern warfare the superpowers felt it necessary to sell them, they responded aggressively to trespassing aircraft. They shot them down when they were in a bad mood and forced them to land when they were in a good mood. Then the interrogations began, whereupon the pilot's fate often depended on how creative he was in answering their questions. Over the years, Globe Aero's pilots had come up with some original excuses, although none as inspired as the one used by Kerby when he was over Egypt, and Cairo Radio advised him that they couldn't find his clearance.

Like any experienced ferry pilot, Kerby tried to confuse the controller with phony numbers, a strategy that might have worked if he didn't have so far to fly before clearing Egyptian airspace. The more Kerby tried to stall, the angrier the controller got.

"We have military fighters in the area," the Egyptian controller warned him. "If you do not land at Cairo immediately, the pilots will be instructed to shoot you down."

"Oh, Jesus! You don't have to do that," answered Kerby. "I always did want to see Cairo. I'm heading for the airport now."

After he landed, several Egyptian soldiers took him to a small room in the bowels of the terminal. An army officer then began asking him why he was flying in Egyptian airspace without a clearance.

"But I got a clearance," Kerby insisted, and dug the document out of his flight bag.

Unfortunately, the clearance had expired several days earlier. Kerby tried to explain that he was delayed because of weather and mechanical problems, which drew little sympathy from the officer, who simply said, "That is no excuse."

Kerby was in trouble. Unless he could think of something to say to this Egyptian, his stay in Cairo might turn out longer than he wanted. Luckily, he remembered reading *The Flying Carpetbaggers* written by the pilot who flew for fugitive financier Robert Vesco. In the book, the pilot told of his experiences flying into Cairo and dealing with Egyptian authorities. Kerby figured he needed to try something equally as simple.

In his most affecting southern drawl, Kerby explained, "The real reason I didn't think President Sadat would get mad at me was because I saw him on TV with President Begin during the peace talks. I thought President Sadat was truly a great man and that Mr. Begin was obviously a jerk. I didn't think Mr. Sadat would get mad, since he said all those nice things about Americans."

The Egyptian officer had not expected this. When he got over his surprise, he asked Kerby to write down what he had said.

"Sure. I'll be pleased to write a report," Kerby said, and proceeded to write three pages of praise of President Sadat. When he was finished, the now smiling Egyptian officer commended him for being such a good judge of character, collected a nominal landing fee, and let him leave.

Unfortunately, not all our problems in Africa were solved this easily. Although we tried to stay clear of the known trouble spots, we weren't always successful. Too many countries were at war with their neighbors or with themselves, so even if you thought you were on the right side, there was no guarantee that you wouldn't be treated with suspicion. Steve Hall, the old

Africa hand I had met during my first trip to the continent, found that out when he was taking an airplane to the Angolan government in Luanda.

Because of fuel problems, Hall had to make an engine-out approach to Cabina, a small uncontrolled airport on Angola's northern coast. Just as he touched down, another airplane landed in the opposite direction. Hall veered off the runway and into the brush, knocking out a couple of flare pots, and swerved back onto the runway after passing the other aircraft, which also swerved off the runway to avoid collision. Six Cuban army officers got out of the airplane and walked over to Hall, who started to say something, but was cut short when the Cuban pilot landed a rifle butt sharply on his shoulder. Hall fell to his knees, at which point the Cuban started swinging at his head.

"Comrade, forgive me," Hall stammered, and went into an idiotic pantomime to suggest he had run out of fuel. Still, the Cuban kept hitting him. Hall broke away from the soldiers long enough to get some papers out of his flight bag showing that he was delivering the airplane to the Angolan government. This is probably the only thing that saved him, because after a long deliberation over his fate, and after they broke his nose, the soldiers decided to let him go.

"It used to be the best stop in black Africa," said Hall, when I ran into him in Abidjan. Over dinner we reminisced about the good old days of colonial Angola. "You could get a first-class hotel, have a good dinner, a bottle of wine and a woman for the night, and still get change back from a fifty-dollar bill. The place is rich with diamonds, iron ore, and petroleum, but now it doesn't have bread in the stores."

"Briggs and Peltz just took two Aero Commanders to Luanda and told me they couldn't get beer in some bar because they weren't Russian," I said. Hall and I were like two old soldiers talking about places we had flown to as if they were campaigns we had fought together. We finished off two bottles of something from a small North African country that had neither the climate nor the soil to produce anything that might be exported under the classification of wine. But it was cheap and flowed freely as we toasted comrades who were no longer with us. For ferry pilots, this was never a subject that could be covered quickly. Whether it was a personal friend or someone we knew by name only, six months didn't go by without another member added to the list. The latest was Gerry Swoboda, the pilot I had met with Hall earlier.

"He was ferrying an airplane from Southeast Asia back to the States when he went down over the Aleutian Islands. But I don't believe he's dead," Hall said, skeptically.

"And why's that?" I asked. "Did they find the airplane?"

"That's just it. They found it floating in the water near Adak, and they found his life raft, too, still packed and lying on the copilot's seat. But they couldn't find the body."

"Maybe he fell into the water before he could inflate the raft."

"Swoboda's not that dumb. He wouldn't have gotten out of the airplane without tying himself to the raft."

"So what happened to him?" I asked.

"You saw all the passports he had. I guarantee he didn't get them without some help. If you ask me, he was recruited by the CIA to fly black missions in Southeast Asia."

Possibly the fact that we crossed international borders with such relative ease had something do to with the rumors that proliferated about ferry pilots, many of them having to do with secret operations undertaken for some intelligence agency or foreign power. One rumor that made the rounds for a while was about a pilot flying to South Africa with a prototype machine gun hidden in his airplane and several South Africans posing as Americans. Another told of a ferry pilot who had some super secret black box in his aircraft that recorded communications when flying near Angola.

Whether or not there was any truth to these rumors is known only to the pilots themselves or the governments that allegedly recruited them. While there might be some importance our mobility could provide the intelligence agencies of the world, for most ferry pilots the significance of troop deployment, orders of battle, or the strength of an enemy's offensive forces paled when compared to the problems we had in just trying to stay alive.

Considering the large number of aircraft we ferried, Globe Aero had an enviable record. Sure, we had our share of accidents and even a few aircraft that were completely wrecked. But, all in all, we were relatively lucky. We lost only one pilot, Walsh—although many of us weren't convinced that he wasn't hiding out in some sleepy Central American bar. Still, none of us had been accused of spying or thrown in prison—until Tom Willett.

Willett was in Abidjan with a single-engine Cessna that was burning about one quart of oil every hour, an amount far in excess of what would keep the engine running long enough to reach Windhoek. Working with a mechanic, Willett found the leak. He was preparing to depart when Kerby—also en route to Windhoek—landed. Willett mentioned the oil consumption problem to Kerby, who, not one to economize his adjectives, questioned the sanity of taking off for a fifteen-hour flight while burning one quart of oil per hour when there were only twelve quarts of oil in the engine.

"I'm not an expert," Kerby told him, "but my steel-trap mind tells me you're gonna run out of oil about three hours short of Windhoek."

"But I found the oil leak. I'm pretty sure it's fixed."

"Oh, you're *pretty sure!* Well, that makes all the difference."

"Come on, Kerby. I'm sure it's fixed. Okay?" Willett said, confident the airplane would indeed make it to Windhoek.

It almost did. He was eleven hours out of Abidjan when the engine began losing oil pressure. Rather than take his chances with ditching in the ocean, he diverted to the closest airstrip, which just happened to be in Angola. Once safely on the ground, Willett was arrested when he told the airport authorities that he was taking the airplane to South Africa. The following morning he was taken to Casa De Reclusao, a two-hundred-year-old prison for political prisoners, and brought before a Soviet colonel who wanted to know what he was doing in Angola.

Willett explained that he was ferrying an airplane to South Africa, but because of engine problems he had to land in Angola. Given Angola's relationship with Africa's white nation, Willett's choice of answers was not an ideal one.

"Who are you flying for?" the colonel asked.

"Globe Aero," said Willett, and then decided that he had a few questions of his own. "Am I under arrest or what? And what have you done with my airplane?"

The colonel didn't answer; instead, he showed Willett a list of pilots who had ferried aircraft to South Africa. Because the South African air force used small airplanes in border skirmishes against the MPLA, any pilot ferrying such a craft was considered by the Soviet advisers to be spies. Willett recognized most of the names on the list as Globe Aero pilots and knew what was coming next.

"You are a spy, and you are not telling me everything. Admit it. You are on a mission for the government in South Africa," accused the colonel and then demanded that Willett tell him what his real mission was.

"You're crazy!" said Willett, who, by now, thought he had listened to enough. "You've got all the answers; you don't need me anymore," he said, and started to leave. He was immediately stopped by the guards, who started beating him. Eventually, they took him to solitary confinement.

Willett's mother had instilled in her son an indomitable spirit, an easy-going manner, and also the Buddhist principles of inner peace. For two weeks, he practiced yoga, meditating, and chanting for hours at a time, right up until the door of the tiny cell was thrown open and he stepped out into bright sunlight.

If the guards were expecting a broken-down shell of a man to walk out, they were mistaken. Willett was laughing and joking with the guards and the other prisoners. He was in as good a mood as he was when he went into solitary—better actually. The other prisoners wanted to know how he managed to come out in such good spirits while others had been dragged out. When Willett explained about Buddhism and yoga, they were skeptical, but they had been in prison for so long that they were willing to try anything. So, in an ex-Portuguese colony in Africa, in a prison run by Russian, Cuban, and Angolan officials, an American pilot taught an assortment of European mercenaries the ancient Buddhist art of yoga.

Meanwhile, back home, Waldman and Willett's relatives were contacting every U.S. and Angolan government official they could think of. Secretary of State Edmund Muskie made a special trip to the United Nations to meet with the Angolan foreign minister.

"You're holding one of our private citizens," the secretary told him. "He is not a government or military person, but a contract, free-lance pilot."

As high-level diplomatic negotiations dragged on in the United States, at the Casa De Reclusao prison in Angola, Willett had the other prisoners meditating, chanting, and making musical instruments for the newly formed prison chorale. Morale among the prisoners had improved to the point that it was undermining the jailers' authority. Six months after he was captured, the Russians handed Willett over to the Belgian Embassy—there was no U.S. Embassy in Angola at that time—and kicked him out of the country.

Willett came back to Lakeland and picked up his life exactly as he had left it. Sadly, however, he died two years later when his fully loaded crop duster crashed while practicing takeoffs and landings.

Globe Aero was basically a small mom-and-pop operation. Everyone was like family—pilots, mechanics, and the office staff, which consisted of McCollum, the office manager; Waldman's wife, Donna, who kept the books, and Holmes's wife, Sierra, who had taken the job of receptionist. Everybody got together for birthdays, Christmas parties, weddings, bar mitzvahs, and now a cremation, requested by Willett's mother in California.

The day after the service, two men from the crematorium walked into Globe Aero's outer office and handed over to Donna a large unmarked box. Inside the box was a smaller, dark, wood-toned, sealed plastic box containing Willett's ashes.

The office was suddenly silent. Nobody knew what to say or do with Willett's ashes so close. Sierra, at her desk next to Donna, was fidgeting in her seat, trying to avoid looking at the box and wishing someone would take it away. McCollum, at his desk on the opposite side of the room, stared at it in mute reverence. Several minutes passed before Donna picked up the box and carried it into Waldman's office.

"What have you got there, a present for me?" asked Waldman, smiling and unaware of the contents of the box.

"No, it's Willett," Donna said, in all seriousness, as she set the box down on her husband's desk and walked out of the office.

Waldman was serious, too—for about two minutes. As he had shown in the past, Waldman had an offbeat sense of humor. He also had these little skeleton's hands in his desk left over from Halloween. Removing the lid from the box, carefully, so as not to disturb the small, sealed box inside with Willett's ashes, he placed the hands so that the fingers were grasped to the outside edge of the box, and put the lid back on. He then went into the outer office. McCollum and Donna were busy at their desks, but Sierra had left the room. When no one was looking, he placed the box on Sierra's desk and walked back into his office. When Sierra returned and sat back down at her desk, the first thing she noticed was the box with the skeletal fingers groping over the edge as if Willett was trying to pull himself out.

"Ahhhh!" she shrieked, jumped out of her chair, and ran from the desk, seeking protection behind McCollum.

"What the . . ." exclaimed McCollum. Then he looked up and backed his chair into Sierra, who screamed again because of the sharp jab to her stomach.

Donna looked up from her desk. "What is it? What's the matter?"

"Th . . . that," stuttered Sierra, her eyes wide with fright as she pointed a trembling finger at the box.

"Oh, my god!" shrieked Donna.

"What's all the commotion about?" asked Waldman, standing at the doorway to his office.

Nobody answered. They were all staring at the box in horror and confusion. Waldman had an idiotic grin on his face, which Donna was the first to notice.

"Phil, you bastard," she said.

"What?" said Waldman, and then started laughing.

McCollum and Sierra, now realizing the appearance of Willett's attempted escape from the box was all Waldman's doing, began laughing. That was all it took. The tension in the room was gone.

"You must be a lot of fun at funerals, Phil," McCollum said.

"We probably shouldn't be laughing," said Donna. "But you know, the person who would've gotten the biggest kick out of this is Willett."

Waldman kept the box in the guest bedroom of his house until Willett's mother arrived for the memorial service. Molly Willett was what one might consider a New Age advocate, whose idea for a service for her son was not so much one of remembrance as it was of communication. Willett's sister, his cousin, and his ex-girlfriend were there, along with several of his coworkers and pilot friends. Twenty-six people, all standing around in Waldman's living room, drinking scotch and mingling as if at a cocktail party, while uncomfortably aware of Willett's ashes resting only a few feet away. Everybody was wondering what was going to happen next, when Willett's mother asked everyone to put down their drinks, form a circle, and link hands.

"Tom may be gone, but his spirit is still with us," she said. "I'm glad all his friends are here, and I welcome anyone to share a story about my son."

She then closed her eyes and looked up at the ceiling. One by one, the gathering voiced a few words about how much they liked Willett, that they would miss him, or that they were sorry for some minor affront. Molly then looked at that person and, interpreting for her son in a faraway voice that echoed from the hereafter, muttered a "That's all right" or a solemn "Goodbye for now." She told everyone that her son heard everything they had to say, and he was pleased.

Willett's sister began strumming a guitar and singing folk songs. Donna read a poem written by Tyler—who was in Africa at the time—that captured the somber feel of the evening:

I chose the skies
that few have known
To follow where
the winds have blown

To battle storms
that none have seen
To find seas of gold
and lands still green

And if someday
I don't return
Don't cry for me
Don't be concerned

For high above
the clouds will sing
For a world I loved
and silent wings.

Willett's death brought a hard reality close to home—ferrying small airplanes was dangerous. People died. And *when* they died, usually in a twisted wreckage of earth, flesh, and metal, a closed casket was almost always common practice. We understood this. Every one of us came into this business knowing that our number could be up at any time. The odds that claimed an average of three ferry pilots each year were tolerable as long as those pilots were faceless strangers. Unfortunately, this failed to provide the sense of noninvolvement that it once did. The Trim God was no longer claiming strangers but pilots we knew, and he was not playing favorites. He took the inexperienced as well as the old timers—Wolf, Swoboda, Martin, and most likely Walsh—no matter how interesting an epitaph we might have fancied. Most recently the Trim God took Bob Iba. Iba was the grandfather of ferrying; he was flying small airplanes over the ocean even before Walt Moody got into the business. Now Iba, too, was dead, shot down over Colombia, by whom and for whatever reasons would forever remain a mystery.

The numbers were closing in on me. I had made 180 crossings, an amount that showed my experience and intimated *some* intelligence. Now, all it did was remind me of how long I had beaten the odds and that each crossing brought me that much closer to the one I would not return from. I felt as if I was living on borrowed time. Ferrying small airplanes over the ocean just wasn't the same.

It didn't help that many of my friends had gotten out of the business and were now starting secure, well-paying careers flying large multiengine airliners. Mantzoros and Gray had gone with Allegheny Airlines, and they were followed several months later by Bennett and Wheeland. Globe Aero's current ranks were also thinning: Peltz went to work for American Airlines and Moriarty had recently headed for Alaska to prospect for gold.

Nor was it comforting that I had recently celebrated my thirty-third birthday and was experiencing the repercussions of passage into my fourth decade, a milestone that made me question where I was going with my life. I liked wandering around the planet like some airborne gypsy, yet I couldn't dismiss a nagging caution that I should be seeking a more conventional career. The now familiar feelings of *yin* and *yang* tugged at my gut. I embarked on each trip in a sort of limbo, my body going through the mechanical motions that had become so routine, yet my mind telling me something was definitely wrong.

I wandered aimlessly to San Francisco, Honolulu, Pago Pago, Bangor, and Gander. The routes were so routine I could tune in the radio without looking up the frequencies. There were the familiar faces of the briefers at the Gander met office and the same repetitious ocean weather patterns I had flown through since I first came here as a novice ferry pilot eight years ago.

Had it been that long? Something was dead inside me, but I had no idea what it was or what caused it. Nor did I know how to fight it even if I could identify the source. Could it be that seeing the death of so many friends was forcing me to come to grips with my own mortality? Was I now frightened of flying over the ocean? Or was I just tired of feeling as if I had to apologize for my country's blunders?

Abroad, America had been suffering one humiliation after another: defeat in Indochina, Soviet advances in Africa and Central America, and the Iranian hostage crisis. The country was shaken by political scandal—not only Watergate, but wrongdoing by the CIA and FBI. Just this morning in Bangor, I woke up and turned on the television to catch the latest news. During the night, six U.S. helicopters had crashed in the desert in an attempt to rescue the hostages being held by Iranian terrorists. I was disillusioned over my country's place in world affairs and a little ashamed to go overseas.

I walked through Gander Airport in a semirobotic state, taking in the surroundings I was so used to when it hit me. Everything was so familiar. I had been down this very same corridor from the weather office to the passenger lounge hundreds of times, but this time was different. This time the airport, the weather office, the snack bar, the Flyers Club, and the hotels all seemed more familiar to me than my own backyard. I was becoming complacent—and bored. I had to make a change and find a new challenge. It was time to walk away from this gypsy life and get a real job in the real world. That decision caused me to miss out on the situation that would develop in the coming months, when Tyler disappeared over Africa and the rumors began circulating that Globe Aero was a front for the CIA.

18

Merchants of Counterinsurgency

Geoff Tyler was a mirthful gremlin who reminded me of Bilbo Baggins in *The Hobbit*. In the short time he had been with Globe Aero, he showed the same knack for getting into adventures as J. R. R. Tolkien's reluctant hobbit. Things just seemed to happen to him, and if there was such a place as Middle Earth, I'm certain Tyler would have come from there. On his first trip, he and Kerby were beaten up by five Samoans in Honolulu, which, as they say, added grist to the mill.

There were all sorts of rumors, most of which had originated from Tyler himself. So they were of questionable authenticity. One had him fighting with the Kurds in Iran as an army intelligence officer—a position for which he still held the rank of captain in the army reserves. Another was the secret missions he flew into the bush whenever he ferried an airplane to South Africa, for which he was paid in gold Krugerrands. Then there was the time he and Kerby were on the Iberian flight from Johannesburg to Madrid, and, in true spy-thriller fashion, Tyler was recruited by a female South African intelligence agent. Kerby swore it actually happened, but we all knew his version of things tended to be swayed by notions of romance and adventure.

Where reality ended and imagination took over, no one could say for certain. Tyler did ferry a lot of airplanes to South Africa, but so did every other pilot when given the opportunity. Now that I had gotten out of the business, Tyler had taken over my unofficial position as the resident African pilot and, as I used to do, broke in the new pilots to this dangerous land. One of those pilots was Raf de Beringer, a Spaniard who grew up in Australia and worked as a flight attendant for British Airways before coming to the United States to fly for Globe Aero. He had taken one trip to Europe when Waldman sent him to South Africa with Tyler leading.

They planned to stop in Washington, D.C., to pick up their visas at the South African Embassy and finish the evening with a drinking tour of the nation's capital. But they never made it that far. While flying the instrument approach into College Park Airport on the outskirts of Washington, de Beringer's engine blew up. He broke out of the clouds at a thousand feet, oil splattering the windshield, and then saw some kids playing in the field he had chosen to land in. At the last minute he veered away from the field, landed between a drainage ditch and a row of houses, and crashed into a telephone pole. The airplane burst into flames.

Tyler had landed ahead of de Beringer and was waiting by his airplane when he heard a loud noise and saw a column of smoke in the direction of the runway a mile away. Someone from the local fixed-base operation (FBO) ran over to Tyler and told him an airplane just crashed.

"You don't know the tail number, do you?" Tyler urgently asked.

When the person told him that he didn't know, Tyler ran inside the airport office to call the control tower and ask them what happened.

"We can't talk about it," the control tower operator said. "We just lost an airplane."

With a car borrowed from the FBO, Tyler raced to where the airplane had gone down. By the time he reached the crash sight, the fire—which had been hot enough to melt the aluminum siding on nearby houses—had been extinguished, and he recognized the wreckage. It was his friend's airplane, but de Beringer himself was nowhere to be found.

A policeman at the scene told Tyler that an air-ambulance helicopter had landed, picked up the pilot, and took him to the hospital. Tyler looked inside the airplane to see what was left of the cockpit, got a whiff of the insulation, and burned his lungs. Then the police took him to the same hospital de Beringer had been taken to. By the time they arrived, Tyler's lungs were inflamed and he was overtaken by a fit of coughing. A doctor sprayed something in Tyler's throat, and he was able to speak again. The first thing he said was, "I want to see my buddy."

"I don't know if you should see him right now. He's under heavy sedation."

"You have to tell me, Doc, what's he like?" Tyler asked, suspecting the worst.

"Not too good, I'm afraid. He's got third-degree burns over eighty-five percent of his body."

Tyler knew the odds of recovering from such extensive burns were not good, but he was persistent about seeing his friend.

"Are you a relative?" the doctor asked.

"No, but I'm the closest he's got here."

An intern fitted Tyler with a surgical gown and mask and led him into de Beringer's room. He did not look good. There was an IV needle hooked up, wires and tubes sticking into him, his nose and ears were burned off, he was blind, and the polyester shirt he had been wearing had melted onto his chest. He was conscious, but Tyler had no doubt his friend wasn't going to live. Still, he tried not to let his voice betray that certainty.

As soon as Tyler spoke, de Beringer recognized him and asked, "Did I miss the kids?"

"Yeah, you missed all the kids."

"Pretty bad landing, though, huh?"

"Yeah," was all Tyler could say.

De Beringer told him about the crash, about crawling over the burning fuel tanks to get out of the airplane, then the wing blowing up, and one of the kids grabbing him while he was on fire and rolling him in the dirt to put out the flames. Tyler stayed until de Beringer's tongue began to swell, and the nurse told him he had to leave so they could hook up the respirator.

"What's going to happen now?" Tyler asked the doctor.

"We're going to put him out for a while."

"Is he going to wake up?"

"Probably not."

Waldman and his wife, Donna, flew to Washington the following day. Both of them were shaken at the sight of de Beringer and felt partly responsible because it was their airplane he was delivering when he crashed. They were also worried about how Tyler was holding up, because he still had to fly over nine thousand miles of ocean. What made it worse was that Tyler's airplane was a clone of de Beringer's—the same make and model with the very next serial number—identical in every way, right down to the color. Tyler wouldn't admit it, but he was scared.

"Look, if you don't want to finish the trip," Waldman told him, "I understand completely. We'll bring somebody else up here. Dawson will be back in the country in a week. He can come and get the airplane and take it on to South Africa."

"No, I don't see any reason to do that," Tyler said. "I'll finish the trip, but I'd like to see them tear down that engine to find out what went wrong."

De Beringer died of a heart attack two days later. Then the vultures came. First, Globe Aero received a bill from the phone company for knocking down the telephone pole. De Beringer's family tried to get their son's body flown to Spain to be buried, but the hospital wouldn't release the body until the bills were paid. They argued, Waldman and his wife argued, but the hospital administrators wouldn't budge. Rules were rules, and they took precedence over death. That's when de Beringer's friends from British Airways got involved.

They sneaked his sister over from Europe by passing her off as a flight attendant and getting her a ride on an airline jump seat. Somehow they got hold of an ambulance and, posing as orderlies, convinced the hospital staff that they were there to take the body to the morgue. As soon as they left the hospital they drove straight to Dulles Airport and a British Airways 747 being readied for the flight to London.

An airliner on the ground between flights is a busy place. There are literally scores of people coming and going—caterers, cleaning crews, mechanics, refuelers, and baggage handlers. Everyone does his or her assigned job with the zeal of a pit crew in the heat of a race. If anyone could move around the airplane unhindered and unquestioned it would be the flight crew—especially the flight attendants. They pretty much controlled the airplane while it was on the ground, so no one thought anything of it when de Beringer's friends—now in their airline uniforms—stowed a large black plastic bag into the cargo bay in a section not likely to be seen or smelled. By the time anyone at the hospital realized the body was not at the morgue and was, in fact, missing, the airplane was over the Atlantic on its way to Europe.

While de Beringer's body was being smuggled out of the country, Tyler was working with the mechanics, a representative of the engine manufacturer, and an inspector from the National Transportation Safety Board to find out why the engine failed. After taking it apart and inspecting every piece, it was concluded that a piston had separated and burst through the cylinder. Tyler wanted to know why. So the inspectors kept looking, until they found a defective part that had broken a fuel line and ultimately severed an oil line, which threw the piston clear out of the airplane. The engine manufacturer's rep knew it, the government inspector knew it, too, but no one would confirm it. The closest anyone came to admitting liability was: "I think we possibly have some reason to believe that we know what might have happened."

With all the lawsuits and multimillion-dollar awards the courts were handing out, Tyler understood why no one was falling all over themselves to admit responsibility. That didn't make him feel any more confident that the same part on *his* airplane wouldn't also break. He asked the mechanics to look at his engine, too.

"And don't just stop at the engine," he said. "Look over the entire airplane."

When the mechanics were finished, Tyler asked if it was safe to fly over the ocean.

"It's a different part," they told him. "The airplanes may have consecutive serial numbers, but the engines don't. They're not even in the same batch."

Convinced that his airplane would not develop the same problem as de Beringer's, Tyler left for South Africa. The airplane showed no problems on the flight across the Atlantic to the Canary Islands or across the Sahara Desert to Abidjan. Tyler was on the two-thousand-mile leg to Windhoek when the electric fuel pump jammed and began flooding the engine with fuel. Still several hundred miles from land, the only thing he could do was turn off the electrical system so the engine wouldn't stall and continue to Windhoek. By morning the coast was covered with fog, and Tyler had no idea where he was. He could turn on the electrical system and get a fix off the Walvis Bay radio beacon, but then he might flood the engine again. Instead, he turned inland. After flying for several hours, he spotted a road between the clouds, spiraled down, and landed.

The only thing Tyler knew for certain was that he was somewhere in southern Africa—close to the Angolan border, but hopefully, on the South West African side. He knew about the fighting in this part of Africa. He figured all he would have to do is start hiking south and sooner or later there was a good chance he would be picked up by South African troops or pro-Western UNITA guerrillas. Then he would just have to hope for the best.

Tyler had no sooner touched down, when he saw them—hundreds of black men in fatigues and brandishing rifles. They ran toward him from both sides of the road and surrounded the airplane as it came to a stop. Tyler knew then where he was—Angola.

A soldier motioned for Tyler to get out of the airplane, which he quickly did after abandoning the idea of turning on the aircraft's emergency transmitter. The last thing he wanted was for the South African air force to find him *and* this company of enemy soldiers. It wouldn't be long afterward that a flight of Mirages soaked the area with rockets and blood. If he was lucky and didn't die, he was sure he would be an object of revenge for the bloodied soldiers who survived. This wasn't his war, and he had no intention of being a hero. As directed, he climbed out of the airplane and raised his hands in surrender.

It was a delicate moment. The aircraft had U.S. registration, and Tyler was white—the wrong color in the wrong part of Africa. With the small American flag on the shoulder of his flight suit, it was obvious that he was not an ally. He heard a voice exclaim "Americano" before the soldiers pushed him to the ground face down and spread-eagled him, jabbering away in Portuguese and local Bantu dialects. Tyler said nothing as they bound his hands behind his back and blindfolded him.

The only thing in Tyler's favor, and probably the only reason the soldiers didn't kill him on the spot, was that he was not military. He wore no uniform, no wings, and no insignia of any kind that could lead anyone to believe he was in Angola on combatant status. He was, however, wearing his Citadel class ring adorned with rifles, flags, and the word *military* in prominent letters. Tyler had to hide it, and there was only one place he could think of. Indicating a pressing need, his captors allowed him a discrete retreat behind a bush, where Tyler slipped the ring from his finger and into a place not likely to be examined.

Six hours later a large Russian biplane landed on the roadway, collected Tyler and a few guards, and flew to the provincial capital of Menongue. He spent the night under guard in a bombed-out house and then was taken aboard a small Russian business jet and flown to Luanda. Tyler was then taken to a private residence, where his interrogation began.

"Why were you spying on Angola?" an Angolan army captain asked. "How long have you worked for the CIA? What have you told the South Africans?"

Tyler wasn't tortured, but it was made clear that he was dead to the rest of the world and that his captors could do with him as they wished. They regulated his meals, which were drugged with amphetamines and tranquilizers so they could control his sleep and his mood swings. They questioned him when he was high, when he was speeding and wanted to talk, and when he was on the bottom, feeling down and depressed, they let him rot. He was told an interrogation would begin in two hours and then he wouldn't see anyone for two days. The questioning might begin at noon or at midnight. And always there was a stranger with the Angolan captain: sometimes a Slav, sometimes a Latin, and at other times, an Oriental.

After six weeks of interrogation, Tyler was taken to Sao Paulo, a dilapidated edifice with high walls and machine gun towers. He found out later that it was known as the "Death Camp." The jailers took Tyler's shoes and belt, told him not to speak with the other prisoners, and shoved him into a solitary eight-by-four-foot cell with walls pocked with bullet holes and rats running across the floor. It was called El Wuntaville Notore—The Last Train. It was death row, the place where they shot people. At dawn, they took prisoners outside, tied them to a wall, and then Tyler heard the guns, sometimes for real, while other times the firing squad shot over the prisoner's head. After six months in solitary, it was Tyler's turn.

As they did with the other prisoners, the soldiers took Tyler outside and put him face up against a wall. He heard rifles being cocked and was certain they were going to execute him. He prayed and waited. Instead of showering him with bullets, the soldiers started laughing, turned him around, and brought him to the main prison block called El Cambio. In Sao Paulo everything was named after trains.

The camp was run by Angolans and Cubans, but controlled by Russians. Tyler suspected they were KGB because as soon as he arrived, they kept after him to make a videotape stating that the United States was supplying UNITA with arms shipments and airplanes.

"We know you have been supplying airplanes to South Africa for the past six months," a Russian officer told him in perfect English.

Actually, Tyler had been flying to South Africa for over a year, but thought better than to correct this Russian. Every time they interrogated Tyler, they told him something different. Now it was his turn. They wanted something from him, so now they could sweat it out. During one interrogation, the Russian, sitting there with a big grin on his face as if he had just won the lottery, showed Tyler a computer printout annotated in Cyrillic.

"Well, Captain Tyler, born twenty-second October nineteen forty-nine," he began, and then proceeded to read Tyler's life history. He mentioned the years Tyler had attended the Citadel, the dates he was commissioned and promoted, his assignments working with the Kurds in Iran and, later, tribes in the Bamian Valley in Afghanistan.

"This is not good," the Russian informed him. "We do not know what you are doing in Angola, but with a background like yours it must be bad. As far as we are concerned, we have enough reason to execute you now."

In the cloak-and-dagger world of spies and counterspies, Tyler was too valuable to be executed as one might an ordinary mercenary. The Angolan government and their Russian benefactors could use him to trade for their own soldiers. The United States didn't have diplomatic relations with Angola at the time, so the Russians leaked word through various embassies that they had an American spy to trade. But they didn't tell Tyler that; they sent him back to his cell, letting him think they could execute him whenever it pleased them.

So Tyler settled into daily life with the four hundred or so other prisoners, mostly guerrillas, journalists, lawyers, and clergymen. Sao Paulo was a political prison, which meant the men incarcerated there had committed crimes against the state, everything from bombing government installations to stealing diamonds. There were Zairians, Portuguese, and South Africans who had managed in some way to offend or threaten the Marxists in power. Among this polyglot of blacks, whites, and browns were several British and American mercenaries.

Tyler's cell mate was Gus Grillo, an ex-marine with arms like pistons and a waxed mustache who had learned combat in Vietnam. Unable to adjust to the sedentary life of a civilian, Grillo came to Angola in 1976 to fight with the FNLA against the communists. Angola had just become independent when the superpowers began plotting like jealous midwives to see who could have the greatest influence over the new country. While the Russians and Cubans came in openly to aid the MPLA, the CIA got in bed with the UNITA and FNLA rebels. As usual, they did it in back alleys, where they recruited the mercenaries and soldiers of fortune to do the dirty work for them.

When Tyler failed to show up in Capetown, the South African air force began searching for him with high-flying reconnaissance aircraft. They eventually spotted his airplane in Angola and passed the information on to the U.S. Department of State. With rumors of Tyler's incarceration filtering in from other embassies, they notified Tyler's mother and Globe Aero. When Kerby found out his friend was still alive, he called Tyler's girlfriend in South Africa—the same woman who was supposed to be an operative for South African intelligence—and volunteered to fly a rescue mission into Angola.

"What're you talking about?" she exclaimed. "This isn't James Bond! We know the Angolans have him, but we don't know where they're holding him. I'd like to do something, but the best thing we can do for Tyler right now is let the diplomats handle this."

Such caution did not stop the press. A magazine in Mozambique, another Marxist African nation, quoted Angolan sources who claimed Globe Aero was a front for the Central Intelligence Agency and branded Tyler as America's premier air pirate. The *New York Daily News* noted that a mercenary pilot with CIA connections had been captured. *Covert Action*, a leftist U.S. newsletter, published an article alleging that Globe Aero was a CIA company and that Tyler was an agent. A reporter from the *Wall Street Journal* came to Lakeland to interview Waldman and snoop out the truth behind Globe Aero's connection to the spy agency.

"The only thing CIA means to us," Waldman told the reporter, "is 'cash in advance.' "

Meanwhile, back in Angola, Tyler was adjusting to the routine of prison life under the tutelage of the veteran mercenary Grillo. "I think they're forgetting about us," the mercenary said to Tyler one day. "We need to do something for political impact."

"Okay, what do you want to do?" Tyler asked, eager to do anything to break the monotony.

"You're gonna have to hang yourself."

"No, let's think of something else," Tyler said, after giving the suggestion about one second of thought.

"We have to. It's the only way," said Grillo, a thoughtful, bullish man with a limp, who had been a prisoner in Angola for five years. He and Tyler shared the cell with another mercenary, Gary Acker, a Californian who sat in his cell and read the Bible for hours, while Tyler and Grillo tried to keep their minds alert by playing chess and philosophizing about the meaning of life. Tyler had grown to respect and admire Grillo's integrity and, in the end, went along with his conspiracy. Fashioning a noose out of a mattress cover, Tyler suspended himself from the top of their bunk in such a way that he wouldn't harm himself, but appeared to be choking to death when the guard looked through the tiny window into the dim cell.

"He's killing himself! He's killing himself!" Grillo yelled, pounding on the door to the cell.

The guard started laughing and hollering at the sight of the dangling American with his tongue hanging out of his mouth. Grillo continued to scream, and the frenzied guard fetched other frenzied guards, who began trying to pick the lock because they couldn't find the key.

"Hurry up, will ya," Grillo beseeched, pointing frantically at the hanging figure of Tyler with his head cocked to one side, his tongue hanging, and his eyes bulging. "Jesus, he's gonna die!"

The guards banged at the lock, frantically yelling at each other, half in Portuguese, half in English. They shot at it with their guns, just like in the movies. Unlike the movies, the lock didn't disintegrate and the door didn't swing open. Instead, the bullets ricocheted around the room, sending the guards screaming and ducking for cover. Only after the guards broke in, by banging on the lock with a pick, did Grillo and Tyler break out laughing.

The prison administrators got the message. Let the Americans get more sun, otherwise they would just make more trouble—especially Grillo, the instigator, who seemed to make a game out of tormenting his captors just to cut down on the boredom. Tyler, on the other hand, usually upbeat and positive, couldn't take his life in prison so lightly. He wasn't a mercenary, and, when he signed on with Globe Aero, he never figured he would end up in an African prison not knowing if he would be shot tomorrow or next week, or rot there the rest of his life.

As bad as prison was, it was worse when he looked up and saw an Angolan Airlines Boeing or Lockheed Hercules flying overhead. Americans were keeping those aircraft flying. Americans were flying those aircraft to and from the diamond camps. Americans were working the oil rigs just offshore. Americans were ready to accept Angola as an honorable trading partner, yet here Tyler was, a fellow American, being held against his will.

It had taken six months for word to reach Washington that Angola was holding an American spy and was going to execute him. The State Department then sent word back to Angola that if Tyler were executed, they would execute a Russian prisoner. That started a complicated trading process involving third party countries and the Italian Embassy in Angola. Before the Italians got involved in any prisoner exchange, however, they wanted to make certain Angola did indeed have someone to trade. They arranged a meeting at the prison between Tyler and the Italian consul, a communist, who hated Tyler from the start.

"What is your name?" the Italian asked.

Tyler told him his name.

"Are you all right?" he asked.

"No. I'm not all right. I want to get out of here."

"Is your health okay?" the Italian continued.

"Yeah, I think so," said Tyler.

Then nothing happened for another six months until late December as Tyler was preparing to spend Christmas in prison. With no warning, he and

Grillo were driven to Luanda Airport and told they were going to be exchanged for two Angolan prisoners. While waiting, word came down that the deal had fallen through, and Tyler and Grillo were returned to prison. Five months later, Tyler and Grillo were paid a visit by the Angolan desk officer of the U.S. Department of State. He told them that officially he was in Angola to participate in discussions regarding South Africa's occupation of Namibia and the withdrawal of Cuban troops from Angola, but had managed to come to the prison to check on Tyler, Grillo, and Acker. He visited them several times during the summer, each time assuring them that negotiations were underway, but that the talks were extremely complicated and could easily come to naught.

Tyler and Grillo waited, but nothing happened. Days turned into weeks, and weeks into months. In early November representatives from the International Red Cross arrived and told Tyler and his cellmates that the exchange was going to take place the following day. As they spoke, word again came down that something had gone awry and the exchange was off. The Red Cross delegates left, promising to return soon, then nothing happened until two weeks later.

"Let's go! It's on," the Red Cross representative told them. They traveled in early morning darkness to Luanda Airport, where two white Twin Otters with large red crosses on their sides were waiting. Tyler, Grillo, and Acker boarded one of the aircraft, along with two Red Cross delegates, a Russian major, and two Dutch pilots, while the bodies of two South Africans were loaded onto the second aircraft.

The aircraft headed for Lusaka, Zambia, where the exchange was to take place. Because Lusaka was eleven hundred miles away, a refueling stop was planned halfway at a jungle airstrip guarded by Marxist guerrillas. The Otters found the airstrip easily enough, but apparently UNITA had found it first. The place was shot to hell, and there was no fuel to be had. The aircraft took off again, this time heading for an old diamond camp on the Zairian border, where they hoped they would be welcome *and* find fuel.

"If we cannot get fuel there, we will have to fly back to Luanda," the pilot told Tyler.

"I don't want to go back to Luanda. I just spent two years there, and this is the closest I've come to getting out."

"I understand how you feel, but we may have to. As long as the Russian has the gun this is his airplane."

Tyler had made up his mind; gun or no gun, fuel or no fuel, he would die in the jungle before going back to prison. He was certain Grillo felt the same. The mercenary was playing chess with the Russian in the back of the airplane when Tyler called him aside and whispered to him the latest development.

"Shit," the mercenary said. "I ain't going back to that goddamn prison."

"That's the way I feel," said Tyler. "Listen. How are you getting along with Boris?"

"Ah, pretty good. He cheats when he thinks I'm not looking, but then when he looks out the window I cheat right back, so I guess we're even."

"Do you think you can take him out?"

Grillo's eyes lit up at the suggestion of physical action. "Yeah. I can break his neck if you want me to. What're you planning?"

"Well, if we can't get any fuel at this camp and the Russian wants to fly back to Luanda, we're going to have to hijack the airplane. But we'll have to kill him first."

The mercenary was silent, thinking the plan through for several seconds before replying, "Okay. You just handle the pilots, and when the time comes I'll take care of the Russian."

"All right, keep your eyes open, and if it looks like everything's falling apart, you know what to do," Tyler said. He started to turn toward the cockpit when Grillo stopped him.

"Wait a minute. What's the sign?" Grillo asked.

"What do you mean, sign?"

"We gotta have a sign so I'll know when to kill him."

"Gus, when the time comes, you'll know it."

"Yeah, but we still gotta have a sign."

"Okay." Tyler gave in. "How 'bout I just yell out, 'Hey, Gus, it's time to kill the Russian.' "

Grillo thought that over. "Tell you what, when the time comes, you just look towards me and scratch your nose, like this," he said, and brought his finger to the side of his nose and rubbed vigorously.

Fortunately, they didn't need to carry out their plan, because the aircraft landed at the mining camp and refueled without incident. During the two hours they were on the ground, however, Tyler fought an irresistible urge to scratch his nose.

Late that afternoon, the two Red Cross Twin Otters landed in Lusaka, where an Aeroflot transport and a USAF C-141 Starlifter waited. Tyler, Grillo, and Acker were escorted to a room in the terminal guarded by a U.S. marine. Several hours later, an Air Botswana C-130 squealed onto the runway and taxied up to the holding area. The rear cargo doors swung open and out walked three Russian airmen, a Cuban soldier, and ninety-four Angolan troops.

The exchange involved the United States, the USSR, South Africa, Angola, Zambia, Cuba, the UNITA guerrillas, and the International Red Cross. Negotiations had taken place on three continents and had taken almost two years to conclude, but finally Tyler was free. He weighed one hundred fifty pounds when he started out on the trip with de Beringer, and now tipped the scales at just over a hundred. As he flew under cover aboard the commercial airliner heading back to the United States, Tyler did something he hadn't done for two years—he began planning his future.

The first thing he wanted to do was celebrate, and get drunk, and stuff himself full of chocolates and pasta and all the things he couldn't get in prison. He would begin to catch up to a world that had been moving along without him for two years. He would get a girlfriend, a car, and an apartment. He might sleep late, or he might get up early to watch a sunrise. He could do whatever he wanted, or nothing if he so chose.

One month later Tyler was back over the Atlantic Ocean in a brand new single-engine Piper bound for Europe.

19

Return to the Fold

Four years in the real world was all I could take. I flew freight for a small cargo airline until it went out of business and then went to work for the government. The aircraft were better, I had all the security and stability I could ask for, and I didn't have to worry about ditching. This was almost the perfect aviation job. Except there was one thing—the bureaucracy. I stuck it out for a while, but eventually it became apparent that if I wanted to stay with the agency I would have to adapt to the system. I decided instead to part company.

Once again I packed my gear and said good-bye to my friends, not even worrying about what I would do next. It was the booming '80s, "Morning in America," according to the latest resident of the White House, and I could do anything I wanted. I chose the one job I liked and I felt was my best destiny, and headed south to Florida.

Walking into Globe Aero's office was like coming home after a long journey to find everything as it was when I left. Phil and Donna Waldman were still there, of course, as was McCollum, at the same desks I had last seen them. But most of the pilots I had flown with were gone. Gray, Bennett,

Mantzoros, and Wheeland were captains with USAir. Holmes was flying rock stars around the country, and Peltz was a first officer with American. Even Briggs, who possessed the right mixture of cynicism, optimism, and humility needed to maintain one's perspective in this dangerous work, had moved to Binghamton, New York, to fly corporate aircraft. Following a secret romance that everyone knew about, he married Kilbourne and took her with him, decimating Globe Aero's cadre of female pilots.

De Long was still around, still living in San Francisco and showing up every now and then to fly his usual leisurely trip. Moriarty, whose resemblance to the rascally Yosemite Sam went beyond mere appearance, was serving time in a Florida prison for smuggling. Tyler was flying for a corporate charter airline in Cincinnati. And while no one had seen or heard from Kuney, his picture turned up in, of all places, *Soldier of Fortune* magazine, walking across an unidentified airport ramp in El Salvador.

Of the many pilots who came to work for Globe Aero during the boom times of the early 1980s, many were now with the airlines, corporate flight departments, the FAA, or, as with one pilot who had followed Moriarty's career a little too closely, got mixed up with drug runners and disappeared into the Federal Witness Protection Program.

The Trim God had been less kind to Finegold. He had realized his goal of flying over every ocean before his twenty-first birthday but was killed when his airplane spun into the ground during a landing at St. Petersburg.

Almost everybody could be traced, except for Rice, who nobody had heard from since Globe Aero moved to Florida thirteen years earlier. Some said he was flying freight out of Miami, and others claimed he was working in a fish market in Pennsylvania. But no one knew for certain. It was as if he had fallen into a black hole and just disappeared.

The bonds of friendship born from trusting one's life to another were strong. Although scattered around the world, many of the pilots called in from time to time to reminisce about the old days or learn the latest gossip of old comrades. There was always talk of a reunion, but the closest we ever came was when Gray, Bennett, Mantzoros, and I met at the annual boat show in Annapolis.

Most ferry pilots had a love/hate relationship with flying over the ocean and quit after several years. As much as we might have liked the independence and freedom, it was pretty much an accepted fact that if you stayed in the business long enough the numbers eventually caught up with you. For those lucky enough to escape this destiny, many found the work becoming routine, and, with the challenge gone, tried to find some new endeavor to take its place. Even Kerby, who had become a legend in the business, quit for a few years, put on a suit, and became a stockbroker. The reason he gave for this career move was typical Kerby: "I looked around and saw more gray-haired stockbrokers than ferry pilots and figured it was time to make a change."

Kerby went into the securities industry the same as he did with ferrying: he studied and researched, drew up charts and graphs, tracked market highs and lows, analyzed and organized, and still it wasn't enough. Kerby needed more excitement, more of the nail-biting, adrenaline rush he used to get from flying through a typhoon in Southeast Asia. He found some of that excitement with options. The only problem was that the company whose options he was trading was under investigation by the Securities and Exchange Commission.

Kerby knew nothing of this investigation or that his buying and selling options in the company triggered a red flag with the SEC, who thought he was tied in with the people manipulating the stock. After an exhaustive probe into Kerby's activities, the investigators concluded that his program trading—although more sophisticated than one would expect from a novice broker—was completely legitimate.

What finally got to Kerby were the little old ladies who came with their pension checks, which he would endeavor to invest with varying degrees of success. After two years, Kerby called it quits and came back to ferrying airplanes. He was now leading me across the Atlantic to make sure I hadn't forgotten anything and to bring me up to date on all the changes.

"Ferrying's a lot different than it was back in the Brown Shoe Days," he said, referring to the 1970s, an era now considered the early days of ferrying. Kerby was a little thinner on top and a bit wider around the waist than I remembered. Other than these few concessions to the greedy demands of time, the years that had passed since I last saw him had not taken too much of a toll. He was still the same opinionated pilot I had frozen with in the

North Atlantic, sweated with in the South Pacific, and gotten drunk with in far too many foreign cities.

"For one thing, most of the airplanes have factory-installed radios. I can't remember when the last time was we had an airplane with temporary radios."

"Well, I can remember," I said. "I lugged enough of them through airports and train stations not to forget. Do we still have to carry around the HFs?"

"Yeah, but now they actually work—not like the old Sunairs we used to have. Phil got rid of them a couple years ago. We got all these Japanese HFs now. Course, we still got the same antenna system."

I thought about all the times I had to crank in and out the one hundred feet of antenna wire and the times I had forgotten to reel it in before landing and then had to untangle and rewind it before I could take off again. Technology, it seemed, had come only so far.

"We can send a vehicle outside the solar system; you'd think we could come up with a decent HF antenna," I said, disappointed over the lack of progress in this one area. It was the only apparent omission.

Ferrying small airplanes, which for years had been accomplished with all the state-of-the-art systems one might find on a papyrus raft, had finally made the leap into the late twentieth century in the most crucial area—navigation. And we owed it all to the Japanese, who took a twenty-year-old navigation system called LORAN, reduced the receivers to the size of a cigar box, and slashed the price to under a thousand dollars. After decades of navigating by guesswork, prayer, and the accuracy of low-level winds, all one had to do now to determine their exact coordinates was push a button.

"You mean when ATC asks for our position we might actually be able to tell them the truth?" I asked, astounded at what in the past had caused a lot of anxiety.

"Yeah, but now you also have to be a lot closer on your navigation. It's not like the old days anymore, when all you had to do was make it to the other side of the ocean and everybody was happy. Now, if you're off course twenty miles, ATC will violate you. Then you have to write letters and go through all this dog-and-pony show with ICAO and the Feds."

Still, that was a small price to pay for the luxury of knowing exactly where I was at any given moment. I didn't have LORAN on this crossing, but as soon as I got back to Lakeland I went to the local electronics shop and bought a receiver to use on my next trip. It was like magic. It was a gift from the gods. Though we didn't realize it at the time, it foreshadowed the end of an era.

Back when the only way to navigate over the ocean was by dead reckoning, there was professionalism and an esprit de corps among the handful of ferry pilots—a confidence gained from challenging the elements and coming out the winner. Now dead reckoning was a lost art. With LORAN, anyone could find his way across the ocean. It not only changed the way the old-timers viewed their work, but also brought a new breed of pilots into the business. They couldn't navigate, knew little about weather systems, and if their LORAN were to suddenly malfunction and they had to find their way across the ocean with only a compass, search-and-rescue crews from North America to Europe would have to be sent out to look for them. They had LORAN and a lot of guts, but little else.

"You can't even go out to dinner with them in a fancy restaurant because they don't bring a sports jacket," complained Kerby. "One guy even came into the office looking for a job wearing jeans." Kerby rolled his eyes toward the ceiling, as he recalled a sight that during Walt Moody's tenure would have been dealt with by a curt lecture on dress followed by a swift boot out the door.

"Times have changed," I said. "Everybody wears jeans now."

"Well, uh . . . maybe so," he stammered for an answer. "But you don't wear 'em when you're asking someone for a job. But that's not the worst of it. This guy had on a belt buckle like I never seen before." Kerby described the large, ornate silver buckle, shaking his head in mortified disbelief and muttering, "Jeesus! Jeesus!"

I didn't know if it was age that had affected Kerby's tolerance level or that he had become less inclined to keep his thoughts to himself. As far as he was concerned, professionalism, a subject upon which he gave little if any ground, was partly reflected in how one dressed. If ferry pilots wanted to be treated like professionals, they should act and dress like professionals. Moody had drummed that precept into our heads so often that not wearing a jacket and tie when with a customer seemed downright slovenly. Now, a jacket and tie were no longer required, and that was just one of the things that exasperated Kerby. Another was Waldman.

For as long as I have known Waldman, he has always had a problem with being completely honest, preferring instead to fabricate whatever story was needed to talk a reluctant pilot into taking a particular trip or to placate an impatient customer who wanted to know where his airplane was. He got away with it in the past, but now he had a more outspoken Kerby to contend with. Kerby had a firm opinion of right and wrong and didn't hesitate to bring what he considered a lapse in ethics to Waldman's attention. The confrontation then played itself out in its usual fashion. Kerby would stand rigid, his hands in his pockets jiggling change, his neck twitching, and his face reddening. While Waldman, sitting calmly at his desk, would tell Kerby or any of the other pilots who disagreed with him that they didn't know what they were talking about.

"You're taking this way too seriously," he would say, followed by his usual assessment. "I think you're having a midlife crisis."

On Waldman's fiftieth birthday, Donna gave him a red convertible. He had an assortment of toys—airplane, motorcycle, antique car—to play with and a melon-sized chunk of flubber on his desk that felt pliable yet firm. I used to knead it and bounce it on the desk whenever I was in Waldman's office until Donna saw me and told me what it was.

"What?" I exclaimed, staring down at the thing, not sure what to do with it.

"A silicone breast implant. Didn't you know that?"

Apart from the eccentricity of using an artificial part of a female's anatomy as a paperweight, Waldman had accomplished a great deal in the twenty years he ran the business. He had built Globe Aero into the largest ferry company in the world and, later, weathered the lean times in the mid-1980s, when the small airplane manufacturing industry went into a slump that drove most of his competitors out of business. Though nothing like the boom years of the late 1970s, ferrying was again picking up and Waldman was taking on new pilots. In addition to his five core pilots, as he called Kerby, Dawson, Egaas, Fryns, and myself, were half a dozen or so freelancers who often flew more than us full-timers did. Waldman, however, sometimes acted as if he were burned out.

"Ferrying probably has five more good years before it just dwindles down to nothing," he confided in me one day.

"So what'll you do then?" I asked.

"That's the scary part. I don't know."

What he was doing now was stashing away as much money as he could before the next down cycle hit. And it looked as if he wanted to accomplish this without the headaches of having to deal with pilots and customers. Toward that end, he was involving himself less and less with the day to day business of ferrying, which he left in the capable hands of McCollum.

Having been with Globe Aero almost from the beginning, McCollum knew the business inside and out. In addition to making sure all the aircraft papers were in order and dealing with the customers, McCollum was now in charge of assigning trips. It was a thankless job, because most of the time he couldn't please anyone. If he had a used airplane or one we felt was too small, he had trouble getting a pilot to fly it. If it sat too long, Waldman and the customer started badgering him and wanting to know when the airplane was going to get moving. In time, he figured out that his job would be much easier if he handled things the same way Waldman and Moody used to.

He and I fell into a routine whenever I went into the office after returning from a trip. I would place a manila envelope with all the paperwork from the trip on his desk, and then say, "Here's the old business, Dave. Do we have any new business?" Then the game would begin.

"We've got an Archer to Kassel, and you're first up," he said, during our latest exchange.

First up meant I was the first pilot back in the country and so had my pick of airplanes if we had several to choose from. Because we had only the Archer—a small, slow, single-engine trainer that I didn't particularly want to take—McCollum figured he had me where he wanted me. If I didn't take the Archer, it might be a while before we got another airplane, a situation he suggested was a strong possibility. This strategy had always worked until I came up with one of my own, which was simply to check with Kerby before going into the office.

No matter where Kerby was, he knew what was happening in the ferry business. He knew which pilots were where, who was flying what, which airplanes Globe Aero was scheduled to get, and when we would be getting them. Having spoken with Kerby, I already knew we would be getting a Caravan for South Africa in a few days. McCollum knew that I knew, but we still had to play the game.

"I thought Phil mentioned something about a Caravan for South Africa," I said, as if this was only a rumor. "Do you know anything about that?"

"Well, let's see," he said, looking over the folders on his desk and wearing an expression that suggested this was the first he had heard of such information.

"Oh, yeah!" he said, suddenly discovering evidence of the upcoming delivery. "We do have a Caravan for Placo. But it won't be ready until the day after tomorrow. You don't really want to wait for it, do you? The Archer's ready now, and I figured you'd want to get going right away."

Choosing between an Archer and a turbine-powered Caravan was easy, especially since the Caravan was going to South Africa. But my answer was also part of the game.

"That's a hard choice," I said, pretending to consider the options thoughtfully. "When does the Caravan have to be there?"

"I told them it'd be there next week," he said. Then, in a too apparent concern for my welfare, he added, "Now I don't want you to have to rush. Maybe you better take the Archer."

We played our game for a while longer, negotiating back and forth, until I settled on the Caravan and he was left with the small, slow Archer to try to unload on the next pilot.

Northern Africa was much the same as I remembered—a quick refueling stop in the Canary Islands, the dead-calm haziness over the Sahara, thunderstorms over Bamako, and the same gnawing feeling that I was someplace where I shouldn't be. Like the old days, I didn't have a visa, so I sneaked into Abidjan through the Air Interivoir hangar and asked the receptionist to call the Hilton.

"Non, non, Monsieur," the young Frenchwoman said. "That is not wise. The Hilton, it is across the bridge, which is closed again today because of the army."

"Why's the army closing the bridge?" I asked. I had read something about political or tribal unrest—these things were normal in most of Africa, but not the Ivory Coast. Of all the ex-French colonies, this country had always been so politically and economically stable that it was often used as a model for other emerging nations to follow. Now, I wasn't so sure.

"Don't tell me the revolution has started?" I said.

"Revolution?" she asked, giving me a confused look. "Ah, non. The army is on strike because the police are paid more money, and so they, too, want a raise. So they close the bridge every day. But they open it again in the late afternoon."

A nine-to-five revolution certainly fit everything I remembered about Africa. "How long's this been going on?" I asked.

"But it is only one week," she said, not looking too concerned. Neither was her boss, a tanned, muscular Frenchman with a droopy handlebar mustache, who looked as if he could've been with the French Foreign Legion and might actually enjoy a good revolution.

"This happens every few years. The soldiers, they get angry, or the police get angry, and they cause trouble for a week. They come to the airport two days ago and say they are taking over. But nobody pays attention. So they take out their guns and they fire into the ground." He described the incident by holding an imaginary gun and pretended to spray the ground with machine-gun fire.

"I'm sure that got everyone's attention."

"Yes, but then nothing happens," he said, with a typical Gallic shrug of indifference. "This is the usual politics. Everything will be back to normal in a few days. But for now, just to be safe, you should stay in a hotel on this side of the bridge."

I ended up at the Palm Beach and easily fell into my old routine of sleep in the afternoon and dinner in the outdoor restaurant by the pool after the sun went down, and it was cool enough to eat outside in comfort. I was halfway through the meal when I heard a familiar voice call out my name, looked up, and saw Clark Woodward standing by my table.

"Boy, am I glad I ran into you," Woodward said, pulling out a chair and sitting down.

I was leery of conversations that began like that, suspecting I was about to be asked to do something that would cost me either money, time, or possibly both. But Woodward was a friend. Although he wasn't flying at the time for Globe Aero, he did work for us on occasion. So when he told me about the problem with his electrical system and asked if he could follow me to South Africa, naturally I agreed.

Four hours after we left Abidjan the following morning en route to Libreville, Woodward's electrical system quit and his radio went dead. I welcomed the quiet. Woodward took two things with him when flying in Africa—not a gun or knife, like some of the pilots carried—that were of questionable practicality. The first was a trumpet, which he played with his microphone keyed during the entire time his radio had been working. And the second, which I saw him carrying after we landed at Libreville, was a roll of toilet paper. Curious, I asked what he planned to do with it.

"You're not serious are you?" he said, giving me a funny look, before offering an explanation that apparently had been well reasoned. "You can't take any chances in Africa. This stuff is like gold here. You never know when you're going to need it, and if you don't have any, God help you, because then it's too late."

Woodward had all sorts of hints to make traveling in the third world easier. The problem was, he shared them with everybody. We came out to the airport the following morning in plenty of time to get an early start, but every time we ran into someone, Woodward struck up a conversation, and after five minutes they would be kidding around as if they had known each other for years. I kept trying to hurry him along, but it was almost noon before we got off the ground.

We flew due south on a course that went straight through Angolan airspace in broad daylight and so close to the MiG base at Mocamedes that they would have no trouble spotting us on radar. This was a definite taboo in the old days, so the closer we got, the more uneasy I felt.

"Isn't this like waving a red flag in front of the gods?" I asked.

"Why? I always come this way."

"Africa must have changed a lot since the last time I was here."

"When was that, nineteen eighty? Eighty-one? That's ancient history. They don't force airplanes down or throw pilots in jail anymore. Angola wants to be friends with the West now. They'll even give you an overflight clearance."

I was skeptical at first, but when Woodward called Radio Luanda to report our position inside Angolan airspace and the controller didn't order us to land immediately or threaten to launch fighters to intercept us, I became a believer.

Windhoek, too, was different, though, the changes were more subtle. It was still a quiet, almost lazy airport, much too large for the few airlines and unscheduled flights that landed there. We parked next to a United Nations Boeing 737, went into the terminal, and waited an hour for customs to show up, stamp our passports, and collect the landing fees. They were expensive even by African standards.

"This is what it's going to be like from now on," said Woodward. "Prices are going to skyrocket and service will go down the toilet. What we saw today is just the beginning. Wait'll after the elections."

"I guess that's what the UN is here for. To monitor the elections and try to make sure everybody votes once and only once."

"What a monumental waste of time that'll be. SWAPO's going to win. Anybody who reads the newspapers can figure that out."

"Then the exodus will start. Just like Rhodesia."

"Zimbabwe," Woodward corrected me. "If you're going to start flying to Africa again, you have to get the names straight."

"How can you get the names straight when they keep getting these disposable governments that change every couple of years? The way things are going, it probably won't be long before we can fly into Angola again and have to stay out of South West Africa."

"The only thing is it's not going to stop there," Woodward said. "Wait 'til it spreads to South Africa after the National Party hands over the government to the ANC."

"It could be worse. They might have to hand it over to the Inkartha." The Inkartha Freedom Party was the militant Zulu faction whose agenda for change embraced the use of tire necklaces filled with gasoline.

"We better hope that never happens. So far the violence has been pretty light and mostly kept to the townships—black against black. But if the wrong party gets in power, it won't be long before it spreads to the white community. Then the Boers will come into the picture, and people will start getting killed by the truckload."

It wasn't only Africa that was changing, but the entire world. Europe had become so congested with airline traffic that we needed slot times to fly just about anywhere. We had to give a month's advance notice to land in Tokyo, and then all we could get was a measly two-hour window. Even Sydney had restrictions against small airplanes landing during peak hours. The days were gone when we could get in an airplane and fly where we wanted when we wanted. Maybe navigation was easier, and maybe we didn't have to cope with beat up, old, temporary radios, but those were the only things easier. Regulations were more restrictive, airline security was tougher, and airplanes had to have a minimum amount of equipment.

The governments in charge of controlling North Atlantic traffic—the Canadian, Irish, Icelandic, and U.S.—had been sending out too many search-and-rescue aircraft to look for small airplanes lost over the ocean. As a result they were beginning to enforce the regulations with considerably more zeal than in the past. Inspectors from the Canadian Transport Ministry were making periodic ramp checks at Gander and St. Johns, and if an airplane didn't have LORAN and HF, they wouldn't give them clearance to leave.

While most of the pilots ferrying small airplanes over the ocean did have LORAN, few of them had an HF radio. Sometimes they didn't even have charts. Instead, pilots hung around the weather office or the refueling shack in Gander or St. Johns and waited for another ferry pilot to come along so they could photocopy his charts. Then they asked if they could cross with him and if he would make position reports for them because their HF wasn't working.

The widespread availability of LORAN had already increased the number of pilots in the business, but then they came out with GPS (global positioning system)—a sophisticated satellite-navigation system that pinpointed one's position anywhere on the planet within an accuracy of thirty meters—and ferry pilots started coming out of the woodwork.

"In the Brown Shoe Days there were maybe four or five ads for ferry companies in *Trade-A-Plane*," Kerby said, as we sat around, griping about the competition. "Now, the last time I looked, I counted twenty-seven."

"And everybody is underbidding us," said Claude Fryns, an expatriate Belgian who always ended his telexes to McCollum with the closure "I am

Claude." Fryns had been flying charter in the Seychelles when he met Kerby on a delivery flight to the remote island. Listening to Kerby's adventurous tales about ferrying airplanes around the world, Fryns decided it had to be more exciting than being on a small island in the middle of the Indian Ocean. A few months later, he moved to Lakeland and pestered Waldman until he finally sent him overseas. Now he was griping like everybody else.

"I just do not see how they can be taking airplanes across so cheap," he complained.

"Because they don't have any equipment," said Kerby. "You just got into the business; you know what you spent. What with charts and radios and all the survival equipment you gotta have, each of us has probably got five or six thousand invested in ferrying. The only thing most of these pilots are doing is getting a GPS—and that's it."

"These guys aren't just saving money on equipment, they're working for nothing," I said. "A lot of them are advertising that they'll fly for expenses just to build flight time. They're taking airplanes to customers we used to go to."

"Well, sure," Kerby agreed, chuckling at the inanity of it. "If you sell at cost and try to make it up in volume, you'll get all the business."

Competition was getting to the point that whenever I went to Gander or St. Johns and saw someone I didn't know hanging around the weather office or the refueling shack, I tried to avoid them. I was usually successful, except for the time the young Englishman I met in the weather office called me at the hotel and started asking questions about how to fly the ocean. Normally, I would've been evasive, but this pilot had a passenger with him—his attractive, South African girlfriend. It was a combination I found difficult to resist.

So I briefed him—and her. I gave them charts and flight plans, explained about weather and how to navigate if their GPS quit. I tried to scare him with stories of ice and ditching and the sun coming up one hundred eighty degrees from where it was supposed to. The English pilot didn't look very confident now. He looked scared to death. Because of the girl, he tried to hide it. They ended up going anyway, but not before drawing up a handwritten will to leave with the refueler . . . just in case.

In a way, I couldn't blame them. They weren't the only ones worried about the ocean. Every time I headed out over the water I had second thoughts, not only because of the weather but also because of the airplanes.

Over the past decade, the cost of product-liability insurance had been steadily creeping upward, forcing many of the small airplane manufacturers out of business. Piper, Beech, and Cessna, the leaders of the industry, were still building airplanes, although on a much smaller scale. And none of them were seeing the profits they had enjoyed in previous years. It was Cessna that was probably hit the hardest though. After losing several major lawsuits, the company shut down its small airplane line and concentrated on high-end turboprops, which most of the distributors or the owners ferried themselves.

This left a vacuum for medium-size, twin-engine airplanes with cabins large enough to stand and walk upright in like we flew so many of in the 1970s. They were still in demand, but because manufacturers weren't meeting this demand and because many of the new airplane distributors had gone out of business, the flying public turned to the used aircraft market.

Used airplanes are like used cars. There are good ones and bad ones. It didn't take long for the supply of good airplanes to be exhausted and more of the latter to start coming our way. We had electrical systems that shorted out and magnetos that backfired. Fuel injectors clogged, gyros started spinning crazily, and turbochargers fizzled up and quit. I broke down so often that I began to feel like Rice when he couldn't go a thousand miles without finding himself beset by some new calamity.

I ferried a single-engine Arrow from Gander direct to the Channel Islands. I had been in the air fourteen hours and had only one hundred miles to go when the engine started backfiring. Power dropped to sixteen hundred rpm, and the engine began shaking so violently, I was afraid it would tear itself from the mounts. It went on like this for two minutes before power came back and everything appeared normal. Five minutes later it started all over again. The cycle repeated itself for the next hour, and there was nothing I could do except maintain altitude for as long as possible and pray the engine stayed in one piece until I reached the airport.

It did, but just barely. The tail and belly of the airplane, relatively clean when I left Gander, were now covered with oil. Further investigation showed less than two quarts still in the engine—the absolute minimum. I learned later that the engine had blown a piston. Whether those last two quarts of oil would have been enough to keep the engine from seizing for another thirty minutes or five was something known only to the Trim God.

Not only were we flying more used airplanes, but most of them were single-engines. Ten years ago, half the airplanes I ferried had two engines, but now, for every twin, I took nine singles. And most of them were small, underpowered trainers that couldn't carry enough fuel to fly the seventeen hundred miles from Gander to Shannon. So we had to take the southern route through the Azores or the northern route through Iceland.

"Oh, Jesus! You don't wanna get caught up in Reykjavik in winter with one of these airplanes. You'll die for sure!" Kerby warned, as we discussed the merits of each route. "Everybody goes through Santa Maria."

"Santa Maria's out of the way," I said. "When you're only doing a hundred knots, an extra four hundred miles can take forever. I don't want to sit in that airplane any longer than I have to."

"You'd like sitting in the water a lot less," advised Jon Egaas. "If you run into problems thirteen hundred miles out, it's nice to have someplace to land. I'm not crazy about flying the extra hours either, but you have to admit, Santa Maria is a lot better alternative than the ocean."

Egaas was an ex-marine corps pilot who, while still on active duty, had landed his combat helicopter on Globe Aero's ramp, aimed the aircraft's cannons at the second floor office windows, and went upstairs to ask Waldman for a job. For the first six months Egaas ferried nothing but crop dusters. Then he figured out how the system worked and began staying away from the office until all McCollum had were big airplanes or turboprops.

"We know you're not crazy about it," I said, both envious and rankled over Egaas's luck in the allotment of airplanes. "But just out of curiosity, when was the last time you flew one of these throwaway airplanes?"

"My union doesn't like me to fly these LBFs," he said with a chuckle, using an acronym that left no doubt as to the collective opinion of these little bitty craft.

Not even Egaas's fictitious union kept him out of small airplanes indefinitely. Winter was fast approaching the North Atlantic and airplanes were getting smaller and slower, prompting the usual grumbling from every corner.

"There's gonna be airplanes scattered up and down the East Coast."

"It'll take longer to get to Europe than to Australia."

"I don't care. I'll sit in Bangor until spring if I have to," said Dawson. "I'm really not looking forward to dealing with all this snow and ice. When are these people going to understand that these airplanes are just too small to take across in the winter? I don't care how careful we are. Sooner or later, someone's going to get up there and kill themselves."

Dawson was a soldier of fortune if ever there was one: combat officer in Vietnam, college student in Germany, and navigation officer on an exploration ship based in Singapore. He had come to work for Globe Aero about a year before I quit, and while I flew only a few trips with him back then, we shared many an all-night card game at the Albatross.

For a brief time, he and I had Gander all to ourselves. When just about every other pilot flying the North Atlantic had made the switch from the Gander-Shannon great circle route to the southern route via the Azores, Dawson and I were the only holdouts. I held out because of nostalgia for the old days when everybody went to Gander, and Dawson because he didn't like change. He took it almost as an insult when I suggested we should also start leaving from St. Johns.

"Oh, no. I've been going to Gander for years, and I don't see any reason to change. If you want to go to St. Johns, go right ahead. But not me. I'm going to Gander. They've got the best weather office in the world, so why would you want to go anywhere else?"

"But all the old guys are gone. Clayton, Doomsday, they've all retired."

"Newhook's still there."

Newhook was a green trainee when I quit ferrying in the early 1980s. Now he was one of the old-timers. Having trained under possibly the best briefers in the world, I trusted his assessment of oceanic weather more than the unknown briefers at St. Johns. So, like Dawson, I kept flying to Gander and went to St. Johns only when forced to because of weather or logistics. But then Newhook transferred out, and I too made the switch.

St. Johns was larger than Gander. It was the capital of Newfoundland and claimed to be the oldest city in North America, with origins dating back to the Vikings. Even the Queen came to visit from time to time—an honor rarely bestowed upon Gander. Its only drawback was the weather. St. Johns rests on a peninsula. During winter, the airport was usually fogged in, and the rest of the year, when storms came up the Atlantic seaboard or swung down from the Great Lakes, they invariably tracked just south of Newfoundland and stalled when they were abeam St. Johns.

Whenever that happened, I usually stayed in Bangor, preferring to wait it out where it wasn't as bleak and dismal. I never had to wait more than two days before the weather cleared and I could fly to St. Johns and leave the next day for Santa Maria. As long as I have been ferrying airplanes, at least once every winter I could count on getting stuck because of weather. It was beginning to look as if now would be that time. I had been in Bangor three days, calling the weather office in St. Johns every day and asking the same question.

"Is it ever going to move?"

"Well, now, it doesn't seem to be doing much of anything, eh," the briefer said. "It's just hanging around south of the island, and we don't expect it to start moving to the east for another three days. You might want to think about going through Goose Bay. They're wide open."

I thought about how many times I had gotten that advice. I didn't mind when it was summer or when I had a good airplane, but not in the small, single-engine Cessna Skyhawk I was flying. The Skyhawk was a two-seat trainer that just barely got up to one hundred knots when it was full of fuel. It was one of the smallest airplanes we ferried, too small to take up in the northern latitudes just to see how much ice it could carry before falling out of the sky.

So I stayed in Bangor, growing more impatient and frustrated each day. Weather over the ocean was good; I just couldn't get to it. Gander and St. Johns, although overcast, were both good. Even the six hundred miles I would have to fly to get to Newfoundland was, for the most part, clear. But it wasn't clouds or ice I was worried about; it was the headwind. With the low-pressure area south of the island, I would be flying directly into a forty-knot headwind. Although I had enough fuel to stay in the air fifteen hours, I calculated that it would take close to ten hours to fly to either of those coastal airports, and I just didn't want to stay in the air that long only to go so short a distance.

The low-pressure area, however, wasn't moving. It just sat there, wrapping around itself, getting deeper, and the winds growing stronger. I waited another day, and, when there was still no change, I left for St. Johns.

I was halfway between Bangor and St. Johns when I ran into a brick wall. My speed over the ground dropped from eighty knots to twenty-seven, a few times dropping even farther to eighteen knots. This couldn't be right, I thought, as I stared at the instrument panel in disbelief. I would have to be flying into a ninety-knot headwind.

A simple time and distance check proved the accuracy of the instruments and that I was indeed crawling over the Earth. The only thing that would cause that was if the low-pressure area had moved closer to St. Johns. I had almost full fuel tanks when I left Bangor, enough to fly sixteen hundred miles in still air, but now, because of the headwind, I would be lucky to make six hundred.

I diverted to Gander, one hundred miles farther from the low, and flew a course that was still into the wind but wouldn't slow me down as much because I would be hitting it at an angle. I also descended to seven thousand feet, the minimum altitude for this part of the route, and where I hoped the winds would be lighter. They were. My groundspeed crept higher and, after a few minutes, steadied at thirty-eight knots. Cars no longer zoomed past me on the highway; now, they gradually overtook me. The clear sky ahead and to the north blended seamlessly with the darkened clouds of the low just to the south in the enveloping twilight. I was fifty miles from Gander when the engine quit.

Instinctively, I reached for the carburetor heat knob, hoping that that was the problem, because oil pressure was normal and I was positive there was still fuel in the tanks. Seconds ticked by as the airplane dropped, giving up precious altitude. There was not enough time to look for a clearing—only to put out a quick call to Gander Approach, more as an advisory than with any hope that they could do anything. I was no more than five or six hundred feet above the ground with trees and boulders rushing up at me when the engine fired, came up to half power, sputtered, and, with a deafening roar, surged back to full power. Gauges danced crazily and then swung over to the right, followed by the nose of the airplane rising until it was above the horizon and pushing upward.

One hour and thirty minutes later I touched down on Gander's east-west runway—not the best landing I had ever made, but I was just glad to be on the ground. Then I noticed that it wasn't asphalt I was on but ice; the entire runway was a solid sheet of smooth, clear ice. The winds were gusting to fifty-five knots, and every time I tried to turn off the runway, the wind caught me, lifted the windward wing, and I started sliding sideways. It took two men steadying the wingtips to walk me to the hangar.

I had been on worse flights, although none that took quite as long to go so short a distance. Yet, this time was different. This time I felt more alone and mortal than I had ever felt in the past. It was not only the ordeal of the flight, but before I left Bangor, I learned that another friend had finally run out of luck and went to the Trim God. The news both saddened me and caused me to wonder how much longer I could continue to cheat death. The pilot had started ferrying airplanes at about the same time I did, and I was the one who had trained him over the oceans. The pilot was Holmes.

After fourteen years of flying the ocean in small airplanes, it was ironic that Holmes's life ended when the plush corporate jet he was flying rammed into the side of a mountain, taking country and western singer Reba McEntire's band with him. Knowing that Holmes had lived on the edge and broken too many rules did little to make me feel the same fate wasn't waiting for me on the next trip. Much as I didn't want to admit it, I knew it was only a matter of time before the odds caught up with me.

Of all the pilots who came to Lock Haven to fly for Walt Moody and the fledgling Globe Aero, only Kerby and I were still in the business. Even Kerby, the guru himself, was flying less and less. That left only me, and I was feeling like a relic from another era.

In the time I had been ferrying airplanes, five presidents had come and gone. I saw the dismantling of apartheid in South Africa and the end of communism in Eastern Europe and the Soviet Union. I watched them tear down the Berlin Wall, and, just two nights ago, I listened to the Red Army Band singing "God Bless America" in Washington, D.C.

The world was moving ahead, and here I was flying over the ocean in airplanes that got smaller each year. Did I choose the wrong path? Should I have stayed in Massachusetts and sought a more conventional livelihood? Or later, should I have tried harder to adapt to the agency in Washington and forgotten about ferrying and this nomadic life of roaming around the planet?

But could I? I wanted to ferry airplanes over the ocean ever since I read Ernest K. Gann's *Fate Is the Hunter*. His wartime recollections of a ragtag group of aviators who ferried oil-guzzling, smoke-belching aircraft through inhospitable skies to hard-to-pronounce names on a map is what had drawn me into this business in the first place.

I always considered myself a pilot whose office was made up of the gauges and switches of an airplane cockpit. I realize now that flying for me was merely a means to an end, and that what I really liked was the travel itself. Ferrying airplanes gave me the opportunity to do that—to heed the call of distant lands nonchalantly mentioned in a ferry pilot's conversation. As Moody had said, we were pioneers whose fortunes were dependent on no one but ourselves—and perhaps the whimsy of an uncertain deity. It was a fanciful if not entirely accurate notion. While there was the explorer in us, there was also the gypsy or the vagabond more at home over the ocean or on the tarmac of another hemisphere with no roots, no ties, and if not a girl in every port, at least a few whose demands and expectations went no further than the present. My friend, Susan Norton, knew that better than I.

"You're C. I., Tony," she had said.

"Oh, no!" I gasped, shrinking back in feigned horror. It sounded as if I had some horrible social disease. "What does it mean? Is it terminal?"

" 'Commitment Impaired,' buddy. And in your case it probably is."

Outside the picture window of the Albatross Hotel bar, snow fell in large flakes that were sucked up by gale winds and then dumped upon mounds that were steadily building and drifting, at times graying out everything except the halo from a dim street lamp.

As the waitress placed another drink before me, I noticed the Globe Aero baggage sticker someone had stuck to the mirror behind the bar years ago. It was faded now, like those for Pan Am and Flying Tiger, almost unrecognizable amidst the newer and shinier ones for Air France, Lufthansa, British Airways, and the countless others that hinted of adventure in faraway places.

The body text cut used for this book is set in Janson Text 10 x 15.
This face was designed in the late 1600's by Nicholas Kis,
a Hungarian in Amsterdam. Just as Kis played with
the thick and thin lines, the ferry pilots played
with the dangers of flying over open
oceans with primitive avionics.